MW00489810

ALSO BY LAURA TRETHEWEY

The Imperiled Ocean

THE DEEPEST MAP

THE DEEPEST MAP

THE HIGH-STAKES RACE TO CHART THE WORLD'S OCEANS

LAURA TRETHEWEY

HARPER WAVE

An Imprint of HarperCollinsPublishers

HarperCollins books may be purchased for educational, business, or sales promotional use. For information, please email the Special Markets Department at SPsales@harpercollins.com.

FIRST EDITION

Designed by Elina Cohen
Art courtesy of Shutterstock / VectorArtist7

Library of Congress Cataloging-in-Publication Data has been applied for.

ISBN 978-0-06-309995-1

23 24 25 26 27 LBC 5 4 3 2 1

TO MY MOTHER
WHOSE LOVE IS DEEPER THAN THE MARIANA TRENCH

CONTENTS

An ocean mapper once told me about a sponge that had shaken up his perspective on surveying the seafloor—not some ordinary dish sponge, but a fantastic deep-sea sponge that is among the oldest life-forms on Earth.

He works aboard an exploration vessel (E/V) called *Nautilus*, where he spends his days exploring the seafloor. Today, around a quarter of the seafloor is mapped,[1] and less than 1 percent has been explored with remotely operated vehicles (ROVs) like the kind found on board *Nautilus*. The ROV is as big as a car and kitted out with sensors, headlamps, and video cameras, which was how Renato Kane first encountered that mind-altering sponge.

Kane has worked aboard *Nautilus* for more than a decade now; he's seen a lot of dives in his time, watching as the ROV picks its way across the seafloor, sampling sediment, discovering new species, capturing incredible new animal behavior, skimming over shimmering hydrothermal vents, and zooming in on creatures hidden in deep-sea corals. Kane is lucky to have the job, and he knows it, but any task becomes routine after a while. Every so often, though, something snaps him out of the everyday and reveals a deeper purpose to exploring the seafloor. Like that sponge he saw at the bottom of the Pacific Ocean.

The off-white sponge that loomed up in front of *Nautilus*'s ROV camera that day was gigantic, rivaling the size of the ROV itself. It must have been hundreds of years old, Kane suspected, maybe even a thousand. Deep-sea sponges grow at a glacial pace. One species

(*Monorhaphis chuni*) lives an estimated eleven thousand years, sprouting spiky appendages 6 feet long. The reason for such astonishing longevity is still unclear, but many deep-sea animals reach an advanced age, so it might be an effect of the dark, cold, sparse conditions on the seafloor. Suddenly Kane saw the sponge's unbelievably long life pass before his eyes. The animal, so often mistaken for a plant, had sat quietly in the same spot for centuries while wars raged, pandemics and prophets came and went, empires rose and fell, and had been in that spot, thousands of feet below the ocean's surface, knowing only dark, cold, still water its entire life. Insulated by the miles of ocean above, the life of a deep-sea sponge is incredibly stable compared to that of humans. It lives in utter darkness, at the same constant pressure and temperature, feeling barely a current in the sluggish waters of the deep—until one day when a big, bright blinking machine sweeps in out of the darkness, peers at it, and moves on, leaving it swaying in the machine's wake. In that brief interaction, we humans are the aliens, visiting another world right here on Earth.

The ROV moved on past the sponge, but Kane couldn't, preoccupied by how unique that moment had been. As the deep-sea explorer Sylvia Earle likes to say, "Drop a stone almost anywhere in the ocean and it's likely to land in a place no human has been."[2]

That dive, and indeed all the dives that Kane had witnessed over his years on *Nautilus*, began to weigh on him heavily. Although all the dives are livestreamed and archived for posterity, the ship would never return to that spot again. "Everything in the moment that you're seeing, it's never seen again," he told me.

At the time, the two of us were sitting in *Nautilus's* data lab as we sailed up the coast of California. The portholes lining the data lab looked like a washing machine on rinse cycle, bubbly waves frothing against the glass and revealing the stormy conditions outside. Otherwise, the room was devoted to screens: computer screens at the half dozen workstations and video monitors covering one wall. Each screen streamed live data from the watery world outside: new maps of the seafloor, new discoveries and data sets that would take experts years to analyze and fully understand. Today, studying the ocean is one of those rare fields where major breakthroughs happen by accident all the time.

And that is the beautiful burden of being an ocean explorer: anything you encounter in the deep sea could be a first.

And there is still so much of it left to explore.

"We know more about the surface of the moon than we do about the bottom of the ocean." This sentence or one like it appears in almost every article you read about the deep sea nowadays. As an ocean journalist who writes and reads exhaustively about the marine world, I've seen it in so many stories that I've lost count. Sometimes the moon is swapped out for Mars or another celestial body, but mostly the statement appears verbatim with little to no explanation. Whenever I read it, I always found myself wondering why: Why do we know so little about the ocean? And why does it make sense that we know more about other planets than we do our own?

When you dig into that sentence a little more, as I did, you learn that it refers only to mapping the seafloor—although our knowledge of deep-sea habitats and history is also paltry compared to our knowledge of land. The best, most complete map we have of the ocean has been made by satellite predictions, and the resolution is so grainy and so general that entire submarine mountains, known as seamounts, are hidden from view. Meanwhile, the topographies of the moon, Mars, Venus, and other celestial bodies are all surveyed at a higher resolution than the global seafloor. As a result of this oversight, the amount of seafloor left to map today is nearly double the size of all the continents on Earth combined.

In 2017, the Nippon Foundation-GEBCO Seabed 2030 Project was formed to finish a complete map of the world's seafloor by the end of the next decade.* Spearheaded by a group of scrappy ocean mappers

---

* I refer to maps and charts interchangeably, but ocean mappers distinguish between them. A nautical chart is a legal document that hydrographers continuously update to reflect the changing conditions in a body of water. A map is a diagram that lays out information spatially.

from all over the world, Seabed 2030 planned to recruit vessels already at sea, from cruise ships to luxury yachts, to crowdsource the map. Its participants wanted to harness new autonomous technology to survey the seafloor with drones. At this pivotal moment in history, when the planet is enduring a series of interconnected crises caused by climate change, Seabed 2030 hopes to finish a map that might help us protect both ourselves and the planet. Never again would we hear that we know the moon better than we do the ocean, and wonder why—because at long last, we were finally going to finish the map.

Intrigued, I started to follow Seabed 2030 and quickly realized that there are good reasons for why we haven't mapped the seafloor. For starters, the ocean is huge. Seawater covers 71 percent of the planet. But it's hard to show people, who live on the other 29 percent, just how big 71 percent truly is. It is vast—incomprehensibly so. The majority of the planet is flat blue ocean, and that's only what we can see on the surface. On average, the ocean is nearly 2.5 miles deep—ten times the height of New York City's Empire State Building.[3] The five ocean basins hold over 324 million cubic miles of salt water, comprising 99 percent of the habitable space on Earth. Most of us will never see even a fraction of this underwater world.

The ocean is also, to put it mildly, an unforgiving work environment. Everything is conspiring against the hydrographer who tries to measure the bottom of the ocean.* You need a diesel-powered survey ship, special expertise, and an expensive deepwater sonar to do so. On the surface, you're battling wind, water, waves, sun, and salt, which are constantly breaking down equipment and testing the personnel; meanwhile, below you is a parallel universe of bone-crushing barometric pressure, freezing temperatures, and all-consuming darkness. During my time aboard *Nautilus*, for example, the weather was so brutal that we managed to map only an area roughly the size of Rhode Island before having to return to land. In places such as the big, wild uncharted

---

* In professional circles, an ocean mapper is more commonly known as a hydrographer: someone who practices the science of surveying and charting bodies of salt and fresh water.

Southern Ocean around Antarctica, roaring waves as tall as skyscrapers are commonplace.

This makes mapping the ocean a very, very expensive endeavor. Seabed 2030 pegged the cost of its mission at between $3 billion and $5 billion. (For context, that's roughly how much it cost to send the *Perseverance* rover to Mars in 2020.)[4] As survey ships move closer to shore, the mapping becomes more complicated—and so does the politics. In fact, mapping contested waters can be a geopolitical minefield. Despite Seabed 2030's clearly stated scientific goals, many nations consider mapping within their territorial waters an infringement on their sovereignty—spying, essentially. There are environmental trade-offs, too. The most detailed seafloor maps are made with top-of-the-line multibeam sonars. From shipping traffic to naval exercises and oil and gas surveying, the increasingly industrialized ocean is a sonic nightmare for whales and other marine life that survive through sound. Should we really be adding more noise to the mix?

I waded into these questions by going on a mapping cruise with E/V *Nautilus*. I interviewed dozens of mappers from all over the world. I attended conferences, lectures, and even an international meeting on naming newly mapped seamounts and canyons. I flew to a remote Arctic hamlet to watch Inuit hunters map uncharted coastlines. I went diving in the Gulf of Mexico and saw seafloor maps lead archaeologists to uncover early human history. I wandered through an airplane hangar near San Francisco filled with ocean-mapping drones.

The history of mapping also raises another troubling question: What happens if we *do* finish Seabed 2030? We know all too well from the seafaring colonizers of the past that a map is not a neutral tool. As the journalist Stephen Hall once wrote, "A map always presages some form of exploitation." Those words followed me to Jamaica, where I watched the governments of the world debate rules and regulations for mining the international seabed. Humanity's mass industrialization of the planet has finally come for the last untouched ecosystem on the planet—would maps pave the way?

One truth did become abundantly clear to me while sitting next to Renato Kane on *Nautilus*: we can map the whole ocean right now. In fact, we've had the tools and the technology to do it for decades. Why haven't we? That question led me back to the sentence that kicked off this whole voyage of discovery: we know the surface of the moon better than we do the bottom of the ocean. It's a cliché, yet it rings true today as a new generation in space exploration dawns. NASA has sunk tens of billions of dollars into the Artemis mission: to return astronauts to the moon and eventually send them to Mars.[5] Meanwhile, we risk leaving behind an unexplored world beneath the waves.

CHAPTER 1

AN EXPEDITION INTO THE DEEP

1.

It was the weirdest job description Cassie Bongiovanni had ever read, and she had read a lot of them lately. With weeks to go until graduating with a master's degree in ocean mapping from the University of New Hampshire, the ambitious, talented twenty-five-year-old was surfing the job boards, looking for her first postgraduation gig. At the time, it seemed as though everyone in her program at UNH's Center for Coastal and Ocean Mapping (CCOM) was going for job interviews and signing contracts.

"To be perfectly honest, I was definitely panicking," she remembered. "That program is so well regarded that almost everyone who graduates gets a job immediately. I didn't want to be that one person who doesn't."

Then an email landed in her inbox, forwarded by a friend of a friend inside the federal government's National Oceanic and Atmospheric Administration (NOAA). They were "looking for some qualified folks" to run a mapping system for 110 days at sea, the email read cryptically. There was no mention of which mapping system might be used or where in the world the mapping might take place—or what the new maps might be used for. Still, it caught Bongiovanni at a moment of peak anxiety about her future.

That was in late 2018, when there should have been no shortage of jobs in the "ocean economy," what the Organisation for Economic Co-operation and Development (OECD) predicted would be a new era of unprecedented growth and investment in the sea.[1] The economic growth flowing from the blue side of the planet would double from $1.5 trillion in 2010 to more than $3 trillion by 2030. That growth would be thanks not only to the traditional titans of the sea—shipping, fishing, oil, and gas—but also to upstarts, such as offshore wind power, aquaculture, marine biotechnology, and seabed mining. The newcomers were infusing the ocean with big money and even bigger dreams. Most of the industries needed ocean mappers such as Cassie Bongiovanni.

It wasn't only economists who were taking a keen interest in the ocean. After years of dwindling public investment and devastating reports on marine health, ocean exploration had suddenly entered a gilded age. Throughout the late aughts and into the early 2010s, a who's who of philanthropic billionaires had founded ocean research outfits and launched state-of-the-art vessels. Ray Dalio, a former co–chief investment officer of the world's largest hedge fund, Bridgewater Associates, had founded the ocean research and media group OceanX. The former Google CEO Eric Schmidt and his wife, Wendy, had created the Schmidt Ocean Institute and launched a research vessel named *Falkor*, after the cocker spaniel–faced dragon in *The NeverEnding Story*. Microsoft cofounder Paul Allen and Salesforce CEO Marc Benioff had founded and funded ocean research departments at West Coast universities. Suddenly the richest people in the world were sinking millions of dollars into ocean research.

The international community got on board, too. In 2017, the United Nations General Assembly declared 2021 to 2030 to be the Decade of Ocean Science for Sustainable Development. Later that year, Seabed 2030 was launched with a mandate to map the entire world's ocean by the close of the next decade—an ambitious goal that had eluded humanity for more than four thousand years. Seabed 2030 declared that it would do it in just thirteen years.

World governments, economists, and jet-setting billionaires had all converged on the ocean just as Cassie Bongiovanni was set to graduate from the Ivy League of ocean-mapping schools. In that new blue economy, she should have had no trouble landing a job. There were some leads, some phone calls, some conversations that seemed as though they *might* be heading somewhere . . . and then nothing. She had started her master's two years earlier, in 2016, when experts had estimated that roughly 15 percent of the world's seafloor was mapped. There was more than enough work for her to do; someone just had to give her a chance.

The 15 percent of mapped seafloor is found mostly close to shore, where governments are required by international law to survey their territorial waters. In international waters, however, the ocean maps are patchy. In the best maps held by Seabed 2030, thin survey lines connect the continents, lighting up a patch of seafloor in an ocean of darkness. Those well-charted routes mostly follow shipping lanes used by international cargo fleets, which carry more than 95 percent of world trade, or they indicate the surveyed seafloor where submarine cables, which carry over 90 percent of the world's internet traffic, lie.[2] In certain places, such as the Gulf of Mexico off Louisiana and the Gulf of Sidra near Libya, the seafloor has been thoroughly sounded for fossil fuels buried beneath the primordial muck. But the majority of the international seafloor is shrouded in darkness: the world's last great mystery.

As Cassie studied the email and its odd call for ocean mappers, she felt the weight of her uncertain future pressing down on her. In just a few weeks, she would defend her thesis. Then, in fairly short order, she would graduate, lose her student medical insurance, move out of her rented student apartment, and probably wind up back at her parents' home in Dallas, feeling like a failure. What did she have to lose? She quickly drafted a reply, attached her résumé, and hit "Send."

Cassie describes herself as shy; she's a planner and an organizer. She has big brown eyes, high cheekbones, and a warm smile. She often pulls her long brown hair back into a practical bun or French braid. At first, she might come across as reserved, even quiet. But very quickly, her goofy sense of humor, along with her whip-smart sensitivity and intelligence, become clear. She's also a millennial through and through: she calls herself a "total derp," cracks jokes about Pokémon GO (which I did not get), and at one point made a reference to the 2001 Julia Stiles movie *Save the Last Dance* (which I did).

"Why are you talking to *me*?" she asked in her typically self-effacing fashion when I first reached out to her. As a journalist, when someone asks me that question, I know I've struck gold. I always prefer the Cassies of the world, the behind-the-scenes types, the people down in the trenches who never think they have a good story to tell. Within them, they always hold some unvarnished insight into the truth.

The thing is, ocean explorers are not known as a modest bunch. That was something else I liked about Cassie Bongiovanni. She does not pretend to be heir apparent to Roald Amundsen, racing to reach the South Pole, or a real-life Captain Nemo.* Nor does she belong to the one-percenters exploring the ocean today, people such as Ray Dalio and the filmmaker James Cameron, who have entire teams to do their bidding. But in 2018, she was on the cusp of doing something extraordinary, something that no one had ever done before in the long, tangled history of humans exploring the sea.

And when Cassie speaks about mapping the seafloor, about seeing something that no one else has seen before, I hear in her voice the ineluctable curiosity that powers exploration.

Growing up in landlocked Dallas, surrounded by fields of pumping oil jacks, Cassie learned to love the earth. That propelled her into a bachelor's degree in geology—a "very common path forward" in oil-happy Texas, she said. During the summer, her family would escape Dallas for the seashore, taking road trips up the East Coast to visit relatives in New Jersey. That was where Cassie had a revelation: she loved the ocean, too, but "I didn't realize you could do anything in the ocean outside of marine biology," she remembered. Then she took an introductory oceanography course in the latter half of working for her bachelor's degree and learned she could study the geology *hidden* by the sea. As she continued to take more and more oceanography courses, she

---

* Captain Nemo is the mysterious creator and pilot of the fictional submarine *Nautilus* in Jules Verne's *Ten Thousand Leagues Under the Sea*, published in 1870. The influence of this Victorian sci-fi classic on ocean exploration cannot be overstated, particularly the fantastic underwater world created inside the submarine. The name "Nautilus" is common in ocean circles, gracing research vessels and deep-sea mining companies alike, while giving a nod to the history and still-unrealized potential of ocean exploration.

realized that mapping the seafloor offered a way to combine her love of the ocean and the earth into a single profession.

The same day she sent off her résumé, her phone rang. A man with a New Zealand accent introduced himself. His name was Rob McCallum. "Let me tell you about the Five Deeps," he began.

2.

While Cassie was on the hunt for a job, Victor Vescovo was looking for his next big challenge. Victor hailed from Dallas, too, and he shared Cassie's Italian roots, but that was where the similarities ended. Victor made his fortune in private equity, and in between stints working for navy intelligence, he has used that fortune to build a modern explorer's life centered around adventure. He has completed the Explorers Grand Slam, summiting the highest mountain on all seven continents and skiing to both poles. He owns a helicopter and a jet that he pilots himself. Yet he looks and sounds nothing like the adrenaline-jacked Texas financier you might expect. He has powder-blue eyes, paper-white skin, and long ash-blond hair that he wears pulled back in a low ponytail. His voice is soft and thoughtful, although prone to nerdy bursts of excitement over military history and science fiction. He seems like someone who has never known boredom for even a moment in his life. Now he wanted to become the first person to dive the deepest point of all five oceans.

The idea hatched slowly. Sometime in the early aughts, when Victor was serving as the interim CEO of one company and the chairman of three others, he began casting around for some new adventure he might like to do.[3] He was aging out of the mountaineering world but still hungered for another challenge—one that required less brawn and more brain. For years, he had followed the record-setting exploits of the British business magnate Richard Branson and his attempts to dive to the deepest point of all five oceans.

In late 2014, Branson abandoned his Five Dives quest after a series of technical failures.[4] One of the more serious flaws? The transparent dome on his specially designed submersible caved in under the simulated pressure of diving miles below the ocean's surface. Some people might see a flattened submersible and think, "You know, maybe diving the world's deepest trench isn't for me." Victor, however, realized that a world record was still up for grabs.

"At that point it was 2014 or 2015, and I was like, 'You gotta be kidding me! Human beings have not been to the bottom of our oceans?'" he said. It was almost personally offensive to Victor that no one had yet explored the deepest parts of the planet. Remarkably, the ocean is largely devoid of deep-sea explorers. The most famous dive took place in 1960, when a two-man team descended nearly 7 miles beneath the surface of the Pacific Ocean, right near the US overseas territory of Guam. At that intersection between two tectonic plates, the old, crusty, dense Pacific Plate pushes underneath the younger, lighter Mariana Plate. That geological roughhousing created the Mariana Trench and the trench's deepest point, the Challenger Deep. The Mariana Trench stretches 2,540 kilometers (1,580 miles)—about twenty times as long as the English Channel—and bottoms out just past 10,000 meters (32,808 feet). Within the already exceptionally deep trench is a little slot that stretches another kilometer deeper: the Challenger Deep, the deepest point in the entire ocean at 10,924 meters (35,840 feet) deep.[5]

The two-man team, comprising US Navy lieutenant Don Walsh and Swiss oceanographer Jacques Piccard, climbed into a basic diving craft called *Trieste*. They descended and spent just twenty minutes at full-ocean depth. Don Walsh later told interviewers that he had expected to set off a wave of exploration in the deep sea. But no one returned to the Challenger Deep for another sixty years.

In 2012, the film director James Cameron dived the Challenger Deep and nabbed the world record for deepest solo dive in the world. But that was not a smooth operation, either. Pressed for time, he dived the deep at night. When he resurfaced hours later, his support ship couldn't find him. A nearby yacht, owned by the billionaire Paul Allen, was called in to help find the floating film director. Various parts of his submersible, the *Deepsea Challenger*, had malfunctioned during the dive—none of the malfunctions life threatening—but the submersible never dived again.[6]

As Victor mulled over making a dive to the bottom of the Mariana

Trench, he found a few things that appealed to him about the mission. He liked the symmetry of reaching both the highest and lowest points on Earth. (That could also net him another world record: first person to scale the highest heights *and* lowest lows on Earth.) Even though he might not speak or look like a Texan, he thinks like one: "I come from that culture in Texas where it's 'If not me, who?' You can't just throw it off on government or somebody else. So it's like, well, 'Okay, why not me?'"

Victor's attraction to diving the deep sea coincided with a wider shift in exploration. Throughout the twentieth century, national governments typically funded scientific or military operations to reach extreme unexplored terrain, but more recently, the world's wealthiest individuals—most of them white men—have been outpacing government investment by forming their own private exploration companies. Critics argue that companies such as Elon Musk's SpaceX and Jeff Bezos's Blue Origin are not a step forward but rather a slide backward, throwing up barriers that only the rich and connected can hope to overcome. By privatizing exploration, such companies make exploration in the twenty-first century look a lot like exploration in the nineteenth century, when the brutal inequalities of England's Industrial Revolution provided enough "gentlemen explorers" with the time and money to pursue a new hobby: striking out for unknown terrain.[7]

Over the coming four years, Victor would sink millions into chasing the five deepest dives that Branson had abandoned. He renamed the quest the Five Deeps Expedition. He bought a research ship. He commissioned a cutting-edge submersible, named *Limiting Factor*, from Triton Submarines in Florida. He hired an expedition leader, the New Zealander Rob McCallum, who brought on a science team that included the deep-sea biologist Alan Jamieson and the marine geologist Heather Stewart.

Pretty quickly, Stewart discovered a major snag in Five Deeps' plan: no one actually knew where the deepest points of the ocean were.[8] "The Pacific is the exception," she explained, because "everyone pretty much knows it's the Mariana Trench, as the Challenger [Deep] has been surveyed a lot." The Southern Ocean, on the other hand, is a massive blank space. Only a fraction of the Southern Ocean has been surveyed with modern equipment, and Stewart estimates that maybe only 1 percent of the South Sandwich Trench—the Southern Ocean's deepest region—has been well charted. "So where's the deepest point in the Southern Ocean? It's like, pick a number," she said with a laugh.

That was where Cassie Bongiovanni would come in, Rob McCallum explained on the call. If she joined the Five Deeps as the lead mapper, her job would be to find the deepest points on the entire planet.

3.

When Cassie got off the phone with McCallum, she was, to put it mildly, ecstatic. "I couldn't sleep that night," she said. "I was so excited about the possibilities. I was daydreaming all night about what my life would be like, if I could go to all of those places while doing the job that I love."

Cassie went to Five Deeps' website, and a sleek world map appeared on the home page: oceans in black, land in blue, the five deeps pinpointed in the Atlantic, Southern, Indian, Pacific, and Arctic oceans. The expedition's first dive would be in the Atlantic's Puerto Rico Trench, with stops in Puerto Rico and Curaçao. From there, the ship would head south to a string of volcanic islets in the Southern Ocean, with stopovers in Uruguay and Grytviken on South Georgia Island, a lonely whaling station where Sir Ernest Shackleton stopped before his doomed trip to Antarctica. The ship would pull into Cape Town, South Africa, before heading east to the Indian Ocean for a dive in the Java Trench, with stops in Indonesia and East Timor along the way. Then it would be on to the US overseas territory of Guam, which is right next door to the Mariana Trench, in the Pacific. The ship would then head north, back to the Atlantic, by threading the needle of the Panama Canal, and head to the fifth and final deep in the Arctic, the Molloy Hole, before finishing the trip in Svalbard, Norway.

If everything went well, a hero's welcome would await the team at the Royal Geographical Society in London, England. In a little over a year, the world tour would not only take Cassie to islands so remote that they nearly fell off the map, and to beautiful bustling tropical cities in the middle of the Indian and Pacific oceans, but also catapult her into the ranks of world explorers. There would be icebergs and penguins, Buddhist monasteries and monkeys, lonely mountaintop castles, beaches covered in billions of fragile cream-colored shells . . . and seafloor.

"Yeah, I was freaking out," she said. "I directed my parents to [the website]. I was like, 'Look at where they're going!'" she said in an excited shriek. Always the planner, she kept a growing list of places she wanted to visit one day. The Five Deeps would cross a lot of them off that list. Victor Vescovo would *pay* her to travel the world and do the mapping she loved to do. Excitement building inside her, she clicked on the website's crew page, and five headshots of older white men appeared, one for every senior position. That gave her pause. In fact, it worried her enough to dampen her excitement, and she wasn't the only one.

At a marine science conference, a scientist had already called out Five Deeps for its lack of diversity, a criticism that would continue to dog the expedition.[9] Back at UNH, Cassie asked her friends and colleagues at the CCOM what they thought about the job offer. The universal reaction from her friends in the mapping program: "You *really* want to go to sea with only guys? Not a single girl mentioned anywhere?" Cassie's concern was mostly professional. Would the older, more experienced men listen to a twenty-five-year-old woman fresh out of university when she spoke up?

There were other red flags. Within seconds of speaking with McCallum, she realized that he was not an expert in the type of ocean mapping Cassie would be doing with Five Deeps. McCallum is a fascinating character in his own right. He's a dive master, a licensed aircraft pilot, and a jack-of-all-trades expedition leader whose role requires him to know every team member's job equally, if not deeply. But he is not a hydrographer, and Cassie clocked that immediately. The first tip-off was the way he referred to the multibeam sonar that the Five Deeps had recently purchased. "He kept calling it 'the sonar' and no one calls it 'the sonar,'" Cassie said.

"It is wrong to call a sonar a sonar?" I asked, puzzled. (I'd been doing the same thing pretty much the entire time we were talking.)

"It's implicit that it's a sonar, because the only way to map [in the ocean] is with sound," she explained patiently. In the ocean-mapping world, people talk more specifically in brand names and model numbers that indicate the sonar's depth and frequency. "I asked [McCallum], 'So do you know what type of sonar you have installed?' He said, 'Kongsberg.' And I was like, 'Okay, so a twelve-kilohertz system?' He said, 'I don't know, I'll just send you the paperwork.'" That worried her. What if she took the job and wound up on a ship in the middle of an ocean, the only person who understood that she was attempting to do something no one had ever done before? The thought was not encouraging.

4.

Cassie's favorite way to show people just how little we know about the seafloor is to open the mapping software on her computer and strip the world map down to what we *do* know about the seafloor. "You can see why this whole mapping thing is important," she said with a laugh as she showed me the process one day over Zoom. "Because there's nothing here! This is why. This is the big picture. There is *no* big picture!" The effect was startling. In an instant, the map went from a rich three-dimensional tapestry of underwater mountains, trenches, and canyons to flat, white nothing. That was especially true in deep waters outside national jurisdiction.

In Cassie's experience, people are often surprised to learn that we haven't mapped the ocean already. It's the twenty-first century, and humanity has done far more impressive things, including landing a robot on Mars and editing human genes. World maps tend to foster the impression that the planet is already charted. When I was a child, I remember running my finger over a spinning globe and feeling the raised bumps that stood in for the Rocky Mountains in North America, the Himalayas in Asia. The ocean, however, was a smooth, blank blue. Back then, it didn't seem odd that the rough contours of land stopped at the water's edge. Perhaps I assumed that the smooth surface indicated water? More likely, I didn't think about it all. But it seems obvious now that the dramatic terrain on land must continue beneath the sea.

It's not easy for Cassie whenever a stranger asks her what she does for a living. She says, "I have to explain it to people: 'The ocean is not mapped.'" That is a genuinely confusing statement, because a quick glance at Google Maps seems to contradict it: the ocean appears to be mapped. Except ocean mappers did not make most of those maps; they've been predicted by satellites[10] circling the Earth, accumulating continuous measurements of the ocean's surface and gravity's pull.

The study of the ocean surface is a "reliable kind of phrenology," writes the science journalist Robert Kunzig, because its dips and bumps hint at what lies on the seafloor.[11] With enough measurements, the satellites can pinpoint a permanent dip or bump in the ocean surface, which means that a canyon or seamount is somewhere in the vicinity. Water naturally piles up on top of seamounts and sinks down into canyons, and the sheer mass of all that water changes the gravity. "People are absolutely mind blown when you tell them the surface of the ocean is not the same everywhere," said Cassie.

If the globe I had as a child had shown the true shape of the planet, it would be one lumpy ball indeed. The first thing that would catch your eye would be the midocean ridge system, an underwater mountain range that runs 40,000 miles around the planet. It is the world's largest geographic feature, but it is mostly invisible to us, covered by a dense cloak of ocean 2.5 miles deep. In just a few places, the midocean ridge rears up on land, as in Iceland, where it has ripped open valleys and thrust tempestuous volcanoes up toward the sky. The true summits of the sea are not the midocean ridges, but the deep ocean trenches that Victor Vescovo intended to dive: the Mariana Trench in the Pacific Ocean, the Puerto Rico Trench in the Atlantic Ocean, the Java Trench in the Indian Ocean, the Molloy Hole in the Arctic Ocean, and the South Sandwich Trench in the Southern Ocean.

The deepest of the trenches, the Mariana, is just shy of 7 miles deep, dwarfing the tallest mountains on land. Mount Everest, for instance, could fit into the Mariana Trench with a mile to spare. The prairies of the abyss are covered in soft sediment as fine and billowy as dust. The abyssal plains make up more than half the Earth's surface and far outstrip all the grasslands of the Eurasian Steppe between Hungary and China.[12] These muddy plains hold what is known poetically as marine snow but is actually the remnants of trillions of dead plankton drifting down over billions of years. There's also eroded rock washing off land, like the never-ending deluge of sediment from the Himalayas streaming into the Indus and Ganges Rivers and out into the Indian Ocean. Two extraordinary fans of sediment sweep off both coasts of India, thousands of miles long and 12 miles deep in some places.[13] The seafloor is also more seismically active than land, with exploding underwater volcanoes, sizzling hot springs, fracturing and rifting tectonic plates, and shuddering earthquakes.[14] The grandest waterfall in the world is not Venezuela's Angel Falls, at 3,212 feet tall; it's on the seafloor between Greenland and Iceland, where cold, dense water from the Nordic Seas collides with the lighter, warmer water of the Irminger Sea and plunges over a hidden cataract 11,500 feet down to the seafloor. Everything about the seafloor is bigger, bolder, and more extreme than the comparatively quiet terrain we know as land.

David Sandwell, a geophysicist at the Scripps Institution of Oceanography in La Jolla, California, is one of the pioneers in mapping the seafloor with satellites. You might expect him to boast about his achievements a little, but he is characteristically blunt about the limitations of his work. "I can tell you what the problem is with the satellite [maps] and that we'll never fix it," he said. "The mean ocean depth is about four kilometers [2.5 miles], and what we're measuring in the satellite is the variations in the pull of gravity. That's the sea surface topography. When you have gravity on the seafloor, it pretty much mimics the actual topography, but as soon as you take [the gravity] from the level of the seafloor up that four kilometers, it gets blurred by physics." This is called upward continuation, and the length of the blurring is equal to the depth of the seafloor. "We're never going to get better resolution than four kilometers, which is pretty terrible," said Sandwell. "We'll never fix it. It's physically impossible."

Over the last twenty years, Sandwell and his collaborator Walter Smith at the National Oceanic and Atmospheric Administration (NOAA) have worked to improve the maps. In 2014, they released a new one that charted the entire global seafloor to a resolution of 6 kilometers (3.75 miles).[15] That was a vast improvement on their earlier map, released in 1997, which detailed the seafloor with a resolution of

12 kilometers (7.5 miles). These are still the best, most complete maps we have of the ocean floor, but they lag far behind the maps we have of the moon, Mars, and Venus.[16] Entire seamounts disappear inside a resolution of 5 kilometers (3 miles), and the true size and location of features in the satellite-predicted maps can be "hugely off, massively off," according to Cassie. She has discovered seamounts miles away from where the satellites predicted. The best way to map the seafloor is to send out a survey ship and sound the seafloor piece by piece by piece.*

Sitting there with her, staring at all the uncharted terrain, I realized that it is not a map. Or at least, it's not what we expect of the maps we see on our phones, with the pinpoint precision of the little blue dot indicating our position. Instead, the best maps we have of the seafloor show us how little we know about the planet. There is so much more down there to explore: blown-out, flattened volcanoes known as guyots, mud volcanoes spewing methane, underwater lakes known as brine pools that are so salty they are lethal to almost every life-form except a few microorganisms that might be analogues to the aliens we seek on distant planets.

Cassie's reverse map also reveals the history of how we've charted the ocean up to this point: secretively, sporadically, often greedily. Traditionally, captains never cared about the seafloor as long as it stayed far enough away from the hulls of their ships. To early navigators, the deep sea was known as simply "off sounding." Translation: deep enough.[17] Our approach hasn't changed that much today. Border control, fishing, tourism, shipping: for most maritime industries, the shape of deep seafloor is irrelevant, off sounding. The exceptions to this rule are the fiber-optic cable and deep-sea mining industries, as well as

---

* Just how long does it take to survey a "piece" of seafloor? That depends on a range of factors, such as the sonar's frequency and the ocean depth. Counterintuitively, shallow coastal waters take longer to survey than deep ocean waters do. Imagine a sonar attached to a ship's hull, pointing down at the seafloor as though it were a flashlight aimed at a wall. The closer the flashlight is to the wall, the tinier the beam of light; the farther away, the bigger the beam. Similarly, in deep water, a sonar expands to capture a greater area in a single pass. In shallow water, a ship has to run many more survey lines to capture the same amount of territory.

the militaries of world powers, which are *very* interested in underwater terrain, particularly within key strategic areas.

If Cassie Bongiovanni were to join the Five Deeps, she would sound the very limits of the planet. Most mappers fresh out of UNH's program at CCOM find stable positions working for the government, the navy, or industry, where they might make maps of busy ports and channels. Five Deeps would take her into the blank spots on the map. The expedition could be the biggest adventure of her life. It could also be a total disaster.

Beyond a slick website, Five Deeps didn't have much to its name. Victor Vescovo was a virtual unknown in the ocean community. The notable names attached to the project provided real-world credibility, but as a team, Five Deeps was untested. And behind the scenes, the expedition was already on the verge of a breakdown. There were legal tussles, power coups, hirings, firings, infighting, million-dollar over-runs, a ship trashed by disgruntled workers, and a near impaling by a flying hook.[18]

In the summer of 2018, months before Rob McCallum reached out to Cassie Bongiovanni, the expedition had intended to set sail for the Arctic Ocean without a mapper on board. The scientists, as well as McCallum and Lahey, were pressing Victor to hire one and install a multibeam sonar on his new ship, *Pressure Drop*. Financially, that was a big ask. The ship had just gone through a refit that had been budgeted at $2.5 million and eventually topped out at more than $12 million.[19] During the refit, a worker claimed to have fallen through an open hatch and settled with Victor for a damage award in the high six figures. Now the team was asking Victor to sink another million or more into a multibeam sonar, plus the cost of hiring an ocean mapper. Victor was wealthy, for sure, but not a Jeff Bezos–level billionaire with unlimited resources or even a James Cameron, who claims to direct some of the most profitable movies of all time in order to fund his deep-sea-diving habit.[20]

"I sold it to Victor as 'You don't want to spend all this money, frankly, and go 'round the world, diving arbitrary points where somebody could come 'round in the coming years who has surveyed these places properly and [say], 'You've dived in the wrong place,'" said Heather Stewart.

According to Stewart, anything less than a multibeam system would raise doubts about the legitimacy of Victor's world record. That's the system that everyone in the ocean-mapping world, from the military to the government to scientists, uses and trusts today.

Eventually, Stewart and the others managed to win Victor over. "He's very goal driven. And he doesn't like to be beaten," Stewart explained. She advised Five Deeps to purchase an EM 122, produced by the Norwegian marine tech company Kongsberg Maritime. "They're tried, tested, that will get the job done for you," she remembered saying. Not long after, she received an email saying that Five Deeps had purchased an EM 124. That was odd. "There is no EM 124," she thought. She reached out to her contacts at Kongsberg, who said there was indeed an EM 124, but it was not yet on the market, nor had it been mounted on a ship before. For Stewart, that immediately set off alarm bells. "I was like, 'You *never* buy the first one. What have you done?'" The EM 124 they purchased even came stamped with the serial number 0001. That might sound impressive, but in reality it was an absolute nightmare because "no one else in the world is finding bugs except for you," explained Cassie. She compared it to buying the very first model of a brand-new iPhone with the absolute latest operating system that no one else had installed.

By that time, Five Deeps had missed the summer weather window to dive the first deep in the Arctic Ocean. In order to make the transit among five ocean trenches in a single year, the expedition would have to pivot around two narrow weather windows at the planet's poles: a January-to-February summer in the Southern Ocean and a July-to-August summer in the Arctic Ocean. But it wasn't the multibeam sonar that caused the delay; it was the submersible from Triton Submarines. The small but scrappy Florida company had made a name for itself building submersibles for clients such as Ray Dalio and his OceanX organization, which had filmed some of the most compelling moments of the BBC's *Blue Planet II* series.[21]

At the beginning of summer 2018, Five Deeps' chief scientist, Alan Jamieson, flew out to the Triton workshop in Vero Beach to see how the build was coming along. "It was amazing 'cause I walked in the hangar of Triton Subs, expecting to see the submarine, and all there was was a titanium ball," he said with a laugh. He had planned to stay two weeks but ended up staying in Florida most of the summer, watching as the Triton team worked insane hours in the swampy heat of the un-air-conditioned hangar. Triton CEO Patrick Lahey, a chipper, relentlessly optimistic Canadian, oversaw the build and held out hope that Five Deeps might make the Arctic weather window until the very end of the summer. It all hinged on a series of test dives in the Bahamas that had to go absolutely perfectly.

On one test dive, Victor flicked on the power for the sub's manipulator arm and a wisp of smoke curled through the cabin. "Do you smell that?" he asked Lahey. "Yes," Lahey responded. Then both men reflexively reached for their scuba regulators, which would give them about a two-minute supply of clean air. They also had a backup supply on board, known as a self-contained breathing apparatus (SCBA), which miners use to survive collapsing seams inside the earth. A fire is a particularly nightmarish scenario underwater, where oxygen is in high demand. The passengers on board the submersible need oxygen to breathe, but the oxygen also feeds the flames inside the tightly enclosed cabin. The smoke inside the submersible cleared not long afterward. It was a false alarm, triggered by a power surge when Victor had gone to use the manipulator arm, but Victor and Lahey had already aborted the dive. Four years earlier, the caved-in dome on Richard Branson's submersible had ended his quest. Triton's titanium construction was sturdier, but the cost overruns and technical glitches were threatening to swamp Five Deeps.[22]

"The sub guys just wanted this damn thing fixed. The expedition guys were trying to herd cats. Victor was like, 'Why isn't this [submersible] ready yet?' not really appreciating the magnitude of what we're trying to build," explained Jamieson. The flaws were all non-life-threatening things such as gonzo alarms on the control board or a surging power supply, but they could still abort a dive at the bottom of the ocean.

By the end of the summer, everyone involved was forced to admit that Five Deeps would not make it to the Arctic in time for the dive. McCallum changed the itinerary, moving the Atlantic Ocean's Puerto Rico Trench to first place; the Southern Ocean slipped into second place, and the Arctic Ocean would come last, in late summer 2019. The four-month delay had cost Victor an estimated $2 million more in crew pay and related expenses. But without a working sub, there was no Five Deeps. There were no science or mapping missions, either. That fall, the EM 124 multibeam sonar would be installed on *Pressure Drop* in Curaçao, a Dutch territory in the Caribbean. Then Five Deeps would set off for a series of final test dives, followed by the first dive in the Puerto Rico Trench, the Atlantic Ocean's deepest point. The fate of the expedition rested on an unproven sub, an unproven team, and an unproven sonar pulling off a dive before the end of 2018—and before Victor's patience and pocketbook wore out.

A few weeks before Christmas, Cassie Bongiovanni touched down in Curaçao as scheduled: the new ocean mapper for Five Deeps reporting for duty.

CHAPTER 2

LOOKING FOR A SHIP

1.

In my interviews with Cassie Bongiovanni, my mind kept snagging on a few key concepts in ocean mapping. Like, how much seafloor is left to map? And why is it so difficult to measure the deepest depth of the ocean?

A quick search online tells me that the surface area of the global ocean covers 139.7 million square miles (362 million square kilometers), or 71 percent of the planet's surface.[1] But what does 139.7 million square miles feel and look like in human terms? There is literally no other space on Earth as big as the global seafloor. That means there is no terrestrial comparison to draw on, such as the height of the Eiffel Tower or the length of Manhattan. Perhaps that's why we stretch for the stars and compare the seafloor to outer space. Perhaps it's why we say things such as "We know more about the surface of the moon than we do the bottom of the planet."

Alan Jamieson absolutely loathes that sentence. One of the main reasons for his vitriol is that it's not *that* impressive to say we know more about a small, dry, lifeless moon than we do about our own big, watery, lively planet. The moon is *tiny* compared to Earth, at about 7.5 percent its size. The North Atlantic alone accounts for more surface area than the entire moon. Even the width of Australia is girthier than the moon's rather slim waistline.[2] In 2019, Seabed 2030 announced that its newest map of the world's seafloor, known as the "global grid," now covered 15 percent of the ocean at the desired resolution.[3] That means we've already mapped nearly one-and-a-half moons' worth of seafloor. "That's pretty good, like why are we hitting ourselves over this?" Jamieson ranted on *The Deep-Sea Podcast*. A journalist even quoted Jamieson as saying the statement in an article. Jamieson, of course, emailed the journalist to clarify that he had never, would never say such a thing and remembered the journalist's response as being "Well, that's what people say. We're going to put it in. This is what people want to hear." The sentence irked him so much that he published an entire paper tracking it down to an academic paper written in the 1950s, when humanity had visited neither the moon nor the deep sea.[4]

In order to understand the enormity of the job facing Seabed 2030, it would be more accurate to say that we'll be mapping *eight more moons' worth of seafloor* by area over the coming decade. And even that comparison doesn't truly do the job justice. The majority of Earth's surface is covered by an average of 2.5 miles of opaque salt water. Water absorbs, refracts, and reflects light, blocking our attempts to map it with lasers and radar, the way we have Mars, Venus, and every other planet without a watery surface. It feels powerful and truthful to compare the ocean and outer space. But when it comes to mapping faraway planets, comparing the seafloor with the moon actually *undersells* the task at hand.

How much ocean is left to map? To answer that question, I called up Tim Kearns, a fast-talking Canadian mapper who runs a nonprofit organization, Map the Gaps, devoted to finishing the world map of the seafloor. Kearns immediately brought up the multilateral search for the missing Malaysia Airlines flight 370, which had created the first detailed maps of a 107,722-square-mile stretch of the poorly mapped southeastern Indian Ocean.[5]

"I saw the data set, and it was wild. It was absolutely beautiful," Kearns said. The new maps revealed seamounts, submarine landslides, and seafloor fractures, as well as two shipwrecks from the nineteenth century. "[The mappers] were out there for a phenomenal amount of time, they collected a phenomenal amount of data, but if you were to plot that on a [world] map, it would be like taking a match-box and putting it on the floor of your kitchen. It was nothing! I'm not trying to disparage it. It was just 'Holy crap, the oceans are really, really big.'"

Though we have been attempting to map the ocean for thousands of years,[6] tragedies at sea, such as the missing Malaysia Airlines flight 370, jolt us into action. The disappearance of Amelia Earhart, the sinking of the *Titanic*, the Indian Ocean tsunami of 2004, which killed a quarter-million people—those macabre events each triggered a surge in studying the seafloor. The surge also comes with a little soul-searching as to why we direct so much money and attention toward exploring distant moons when our own planet is so poorly understood. The National Oceanic and Atmospheric Administration has been surveying the United States' coastlines for more than two hundred years, but NOAA didn't have a department dedicated to ocean exploration until 2001. NOAA's budget, which includes funding for ocean mapping, is about a fifth of what is spent on exploring outer space. NASA's generous budget rises each year, while government support for ocean research has either declined or plateaued. In 2021, NOAA's budget totaled $5.4 billion, a 1.4 percent increase over the past year.[7] In comparison, NASA's 2021 budget reached $25.2 billion, an increase of 12 percent over the year prior.[8] Suddenly, in tragedy, we grasp just how small and vulnerable we are on this big planet we know so little about. But historically, our interest in the deep sea tends to fade, and we move on to other pursuits.

Another concept I kept snagging on in my conversations with Cassie: Why do ocean mappers struggle to measure what seems like a very finite thing, the bottom of the seafloor? Even with the best multibeam sonars and mappers, the depth measurements of the deep sea often come with a significant margin of error, as much as fifty feet in both directions.

"That is my biggest frustration," Cassie said with a sigh when I asked her that question. "I have explained it until I don't even have words anymore. It's the uncertainty of where our technology is now. I cannot discern with one hundred percent certainty that this one-meter spot [on the seafloor] right here is deeper than that two-meter spot right next to it. I just can't. We are not able to have that kind of resolution in these deep waters with our current technology. Impossible."

"Right, right," I said. "Of course." And then a moment later: "But, like, *why* can't we know that?" And Cassie, God bless her, launched into yet another explanation: "I equate it to taking a laser pointer and, say, if you're standing at the bottom of Everest at sea level, pointing a laser pointer to the top and expecting it to be within a centimeter accurate, that's what you're asking me to do." That would be tough, I agreed.

By now it had become obvious that I would need to go to sea myself if I were ever to understand just how big the ocean truly is and how hard it is to map the seafloor. Because the ocean is mapped through sound, the seafloor is "heard" rather than seen. But I'm still human. I have to see something to believe it.

2.

Many of the ships surveying the seafloor today are run either by the military or by industry. Neither would have much interest in allowing a nosy journalist like me on board, who would ask all kinds of inconvenient questions about what they were mapping on the seafloor and why. That left a handful of research ships, run by the federal government, universities, and a few philanthropic and nonprofit organizations. The option nearest to my house was a fifteen-minute drive on the I-5 to San Diego Harbor, where the Scripps Institution of Oceanography tied up its fleet of research ships. There was also NOAA's fleet of research and survey vessels, that of the Woods Hole Oceanographic Institution (WHOI) in Massachusetts, and the Schmidt Ocean Institute and its exploration vessel, *Falkor*.

And so began the grunt work of sending emails, phoning scientists, and asking for a berth on a mapping cruise. At nearly any other point over the last twenty years, that would have been relatively easy. Scientists, like writers, are generally happy to tell you all about their work. During the height of the COVID-19 pandemic, however, that became almost impossible to do in person. The US fleet of research vessels had adopted strict protocols for keeping the disease off its ships. As we learned from the COVID-19 outbreaks on cruise ships in the early days of the pandemic, viruses spread extremely well on ships. With its tight hallways, handrails touched by everyone on board, and big enclosed dining halls, a ship is an incubator of illness. In response, US research vessels cut their ship staff down to the bare minimum of people needed to operate the vessel and conduct experiments. With so many researchers desperate to gather data, there was not a single berth to spare. "In any other year, I'd be happy to take you, but . . . ," began many apologetic responses from researchers otherwise eager to help me.

An expedition leader with the Ocean Exploration Trust (OET) told me much the same. OET operates the exploration vessel *Nautilus*, which broadcasts all its on-board science via a YouTube channel. During the summer of 2020, it surveyed along the Pacific Coast of North America, filling in maps in the gaps of the United States' Exclusive Economic Zone (EEZ). Every coastal nation controls the 12 nautical miles directly offshore, but it also has the exclusive right to the maritime resources in its EEZ—a vast marine territory that extends another 200 nautical miles offshore. The United States' EEZ is over 3 million square miles, a surface area greater than that of all fifty states combined.[9] The majority of it is still uncharted.

Over the coming months, as I tried and failed to find a ship to take me to sea, I watched the E/V *Nautilus* live feed avidly. The crew surveyed a canyon that supported a rich marine ecosystem that the government was considering listing for national protection status. The ship returned to a bubbling plume of methane seeps to understand the chemosynthetic communities that thrive there. The crew scoured the seafloor for fragments of a fallen meteorite and investigated the carcass of a dead whale and the scavenging community that had grown up around it.

Watching *Nautilus* explore online was helpful, but it was not quite close enough. At times, I wanted to sit down beside the ocean mapper and experience it all myself. I kept emailing and calling other organizations and researchers, kept watching E/V *Nautilus* prowl along the

West Coast, and waited for the global pandemic to subside. A year later, just as I was starting to despair that I might never get an inside look at mapping the seafloor, the expedition leader aboard *Nautilus* called me back. She had an empty berth on a mapping expedition that would be leaving in just over a month. Did I want it? Yes, I said. Eagerly.

3.

Something shuddered beneath my sleeping body, waking me with a start. "Where am I?" I wondered aloud, looking around at the unfamiliar room. I was in a single bed, the ceiling about a foot from my raised head, curtains hemming me on three sides. "Riiight," I remembered, flopping back onto the bunk bed. I had fallen asleep on *Nautilus* the previous night; now, at 6:00 a.m., the ship was pulling away from the Port of Los Angeles, right on schedule. Another shudder rose from the engine room deep inside the ship, rattling my berth once again.

The previous afternoon, right after boarding, the expedition leader, Nicole Raineault, had given me a tour around *Nautilus*. The 68.23-meter (224-foot) vessel is almost identical to Victor Vescovo's DSSV *Pressure Drop*: they're similar in size, carry a similar number of crew, have the same steel A-frame on the stern used for plopping scientific toys into the ocean. Both have a Kongsberg multibeam mapping system installed. On the top deck of both ships, the superstructures are studded with radar and satellite hardware, identifying them as research vessels.

As Raineault took me through a wet lab, a production studio, a mapping room, a hangar, a workshop, and onto the bridge, where the captain piloted the ship, I felt as if I'd suddenly dropped into my all-time favorite movie, *The Life Aquatic with Steve Zissou*. Specifically, into the scene where Zissou, played by Bill Murray, shows the audience his ship, *Belafonte*, in a cross-section cutaway. "Do you ever feel like you're in *Life Aquatic*?" I asked Raineault cautiously. She laughed immediately. "It's almost too close to real life," she replied. It didn't take long to figure out who Raineault was on board: Eleanor,

Steve Zissou's wife, played by Anjelica Huston, who keeps the whole operation afloat.

Down in the data lab, Raineault introduced me to the three mappers who would be on our weeklong cruise to Oregon. The ship would thread through a notoriously rough patch of water off the coast of California, where the crew often struggled to survey the seafloor. Trading off eight-hour shifts, the mappers would oversee *Nautilus*'s EM 302 multibeam sonar.* The EM 302 is considerably older than the EM 124 aboard *Pressure Drop*, but it still cost a million dollars when it was installed back in late 2012.† Over its tenure, the EM 302 on *Nautilus* has mapped more than 870,000 square kilometers (336,000 square miles) in the Caribbean Sea, the Gulf of Mexico, and the Pacific. At the end of each expedition, the new maps are fed into a network of databases and eventually land in Seabed 2030's growing global map.

One mapper asked me about my book. "It's about mapping the seafloor and Seabed 2030. It starts with the Five Deeps Expedition," I started to explain. She listened politely before telling me just how frustrating she found all the attention focused on Five Deeps. Ocean mappers like her had gone to sea for years, quietly putting the puzzle pieces together with nearly no recognition. Then this rich white guy comes along and the media eats it up.

"Well," I clarified, "it focuses more on the lead mapper with Five Deeps."

"That's a little better."

---

* In early 2023, OET upgraded its sonar system, installing a new Kongsberg Simrad EC150-3C 150 kilohertz transducer on E/V *Nautilus*.

† The numbers in the sonar's model name refer to frequency: the EM 302 operates at the higher 30-kHz wavelength, whereas the EM 124 operates at the lower 12-kHz wavelength. Higher-frequency sonars such as the EM 302 can survey seafloor as deep as 7,000 meters (23,000 feet), which covers most of the world ocean. A lower-frequency multibeam such as the EM 124 can reach even deeper, to the absolute bottom of the ocean, just past 10,000 meters (32,808 feet). However, lower frequencies are more harmful because the sound travels farther and risks disrupting marine mammals.

I had heard that complaint from ocean mappers before. As much as they appreciated a multimillionaire bringing money and publicity to the cause, the congratulatory coverage was a *bit* much. It wasn't as though a financier had *invented* ocean mapping or something. It reminded me a little of the debates about neighborhood gentrification: Scientists had labored for years on gritty ships, pulling twelve-hour workdays, away from their families for weeks at a time, in order to haul back some seemingly small but scientifically priceless piece of data about the deep sea. Then the billionaires, the new kids on the block, had strolled in, taken a look around, and decided that the neighborhood needed a bit of sprucing up. It's not as though the money for ocean research wasn't needed or appreciated, but some mappers questioned the intentions of the rich, powerful newcomers. Would they dictate the kinds of research questions that might be asked? Was all the philanthropy just a way to burnish their legacy? In the early days of Five Deeps, the press often portrayed Victor as a record chaser rather than a serious ocean explorer. One story that made the rounds involved his asking Triton to make a marblelike submersible without a manipulator arm or even a window to see the ocean floor—an infuriating story for ocean scientists, who desperately want that deep-sea data. (Victor insisted that it had never been a serious suggestion.)

OET and *Nautilus* had a proven track record and hard-earned credibility in the ocean community. Founded by the world-renowned oceanographer Robert Ballard in 2008, OET worked with the government and universities to conduct research that no one else in the United States was doing. Bob Ballard is a living legend, up there with other ocean luminaries such as Jacques Cousteau and Sylvia Earle. In 1977, he was part of the team that discovered the first hydrothermal vents at the bottom of the Pacific Ocean. Up until then, scientists thought all life on Earth was photosynthetic and dependent on the sun for energy. When the team of marine geologists, geochemists, and geophysicists stumbled upon mussels, clams, and crabs living off the minerals spewing from underwater hot springs, there was not a single biologist on board to interpret the findings because no one had expected to find life on the seafloor.[10] Those vents, or ones similar to them, are suspected to be the birthplace of all life on Earth. But Ballard is perhaps best known for his part in discovering the wreckage of the *Titanic* in 1985.

Though it was cast as a purely scientific expedition at the time,[11] more than thirty years later Bob Ballard revealed that the search for the *Titanic* had been a great cover story for what had been primarily a military operation. Funded by the US Navy, Ballard was sent to investigate two sunken submarines, both treasure troves of nuclear military technology, without tipping off the Soviets. Any remaining time on the trip he would be allowed to spend on discovering the most storied shipwreck of all time. It's only one recent example in a long line of science benefiting from tactically relevant expeditions at sea. Ballard, meanwhile, still can't talk about his other navy missions. "They have yet to be declassified," he told CNN in 2018.[12]

Compared to the research vessels run by billionaire-funded philanthropic organizations, Ballard's *Nautilus* takes a more old-school approach. The science crew is referred to as the "Corps of Exploration." Everyone is encouraged to wear the navy-blue *Nautilus* hats, vests, and shirts emblazoned with a compass. No alcohol is allowed. The food is good but not fancy. Gluten-intolerant eaters can bring their own food. At any time of the day or night, you can go to the *Nautilus* website and watch a split screen of the ship, scientists pipetting in laboratories, or the crew wrestling an ROV out of the sea.

After my tour with Raineault, I went for a walk along the top deck and came upon an open manhole that looked straight down into the engine room below. I stuck my head down into the warm darkness for a moment and pulled it back out instantly. It was hot and noisy down there, like a dragon's den, and that was *before* the engines were turned on full blast. Now I was lying in my top bunk, gently rocking from side to side as those engines took us out to sea. Outside my cabin door, I heard chairs being pushed back against a wooden floor, cutlery clinking in the sink. Early risers were moving around the mess just outside my door. I whipped back the curtains on my bunk, and a huge grin spread across my face. Through the cabin porthole, I could see the cranes of the Los Angeles megaport melting into the overcast horizon.

4.

Every morning on *Nautilus*, expedition leader Nicole Raineault scrawled the day's objectives on a whiteboard in the galley. For two days, she wrote simply, "Mapping. If conditions allow." Conditions did not allow. Immediately after we exited the protected channel off Santa Barbara, fierce open-ocean headwinds slammed into *Nautilus*. Swells of 3 to 6 feet rose past 9, 10, 13 feet high. The ship's bow tipped high into the air over each wave. The winds ratcheted up to 33 knots (38 miles per hour). The ocean was dotted with foamy whitecaps as far as the eye could see. On the Beaufort scale, a mariner's measure of sea conditions, they were near gale force. On land, it would feel like walking into a heavy storm as your umbrella is ripped from your hands.

Up ahead was Point Conception, a natural boundary between northern and southern California. North of the point, toward San Francisco and the Bay Area, the land is damper, foggier, and taller, with dense forest. South of the point, toward Los Angeles and San Diego, the land is drier, warmer, and covered with scrubby deserts and chaparral. It's a dividing line out at sea, too, where ocean currents collide like the upward drafts of wind buffeting the edge of a cliff. It's "the largest point on the coast," wrote Richard Henry Dana, Jr., in *Two Years Before the Mast*, an 1840s sailor's memoir that had introduced Americans to California back when it still belonged to Mexico. "An uninhabited headland, stretching out into the Pacific, and [it] has the reputation of being very windy. Any vessel does well which gets by it without a gale."

As we rounded the point, the seas grew taller, rowdier, and rougher. Suddenly I had a newfound appreciation of the phrase "smooth sailing." Out there in the open ocean, it felt like riding a roller coaster gone rogue. Everyday tasks, such as showering and walking, became a challenge. The shower in my cabin was one of those standing-closet affairs fenced in by transparent plastic walls. With one hand against the side, I lathered up quickly with the other hand, feeling a wave rising beneath the ship. Then I stopped soaping and held on for the impact: a moment of weightlessness, followed by a steep drop and a slam back down to the surface. Then I rinsed, bracing for the next slam. Whenever

I felt nausea rise in my stomach, I looked down at the water sloshing and churning around my feet in the bottom of the shower, using it as a gimbal inside the windowless room. After I carefully extricated myself from the shower, I dressed as the ship rocked back and forth. A lean over to port, then a lean over to starboard; the rhythm repeated with endless variations and syncopations.

Outside the cabin, I walked, or rather barreled, down *Nautilus's* tight hallways. At the foot of a stairwell, I crossed paths with Renato Kane, an ocean mapper. "This your first time at sea?" he asked as we held on to a door frame, the stairwell shuddering around us. Kane went to sea so often with *Nautilus* that he didn't keep a permanent home on land anymore. I nodded. He raised an eyebrow. "This is . . . pretty rough for a first-timer."

The ship's crew, a tough and intense group of Ukrainians, seemed impervious to the conditions. As the winds ramped up, the head chef, Anatoliy, blasted AC/DC's "Thunderstruck" in the galley. "He does that every time the weather gets bad," Raineault said with a chuckle as she scrawled the day's activities on the whiteboard: no mapping again.

Later that day, I walked past the gym on the lower deck and glimpsed a crew member, zipped up head to toe in a gray sweat suit, running full speed on a treadmill as the room tilted back and forth at a nearly forty-five-degree angle. Just the sight of it made me blanch, and I hurried away to get some fresh air. A few seasick crew members had taken up a silent vigil on the protected back deck, looking out at the ocean as they tried to calm their stomachs.

The cruise had its sights set on a poorly surveyed stretch off the California coast. In 2019, the Trump administration had thrown its support behind ocean mapping in a presidential memorandum directing the government to create a national strategy for charting the uncharted US EEZ,[13] mainly along the remote coastlines of Alaska. NOAA had also released a map layer that tracked gaps in the map. Research ships such as *Nautilus* layered the NOAA map over their route through US waters, trying to hit as many gaps as possible.

Whereas land surveyors use the speed of light to map the rise and fall in the terrain, ocean mappers use the speed of sound, like a bat reverberating clicks off the walls of a dark cave. All sonar, from the most basic fish finder to the powerful multibeam bolted to the hulls of E/V *Nautilus* and DSSV *Pressure Drop*, follows the same principle. The sonar emits a ping that travels through the water column, bounces off the seafloor, and returns to the sonar. One second of travel time is roughly equal to one mile. Divide the ping's travel time by two, multiply it by the velocity of sound in water and—voilà!—you have a measurement of the depth. Sounds easy, right? Except that the water's salinity, temperature, and pressure, along with a slew of other factors, warp the ping as it travels through the ocean.

Rough weather also complicates the drawing of an accurate, detailed map of the seafloor. The mapping room, also known as the data lab on *Nautilus*, is one of the worst places in which to sit in high seas. Situated at the near center of the ship, the open-plan office has very little natural light or fresh air. Three of the four walls are covered in computer screens, streaming live data from all over the ship, while the fourth wall has a tiny row of portholes that are often frothed up by the waves and foam outside.

There are various types of sonars on board *Nautilus*, and each "hears" the seafloor in a distinct way. The ship crew on the bridge operates a single-beam sonar that sends a single, straightforward ping down to the bottom and back. The single beam reveals a large snapshot of what is immediately beneath the ship, and that's good enough for the crew, who want to avoid running aground. True to its name, the single beam sends a single ping, down-up, down-up, down-up, across the seafloor. The multibeam, on the other hand, emits hundreds of pings simultaneously, raining down into the deep like a huge fan of sound.

It works a little like this: Imagine you had to figure out what an unknown object was in the dark, but you could use only one finger. It would take a while to trace its shape—that's single-beam sonar. Now imagine you could pick up that object with your whole hand. You would know what you were holding far more quickly. That's what a multibeam does, and it's why ocean mappers prefer using it for seafloor surveys: it bathes the seafloor in sound, capturing a clean snapshot of the underwater terrain.

Down in the data lab, I watched new seafloor maps stream on the main computer screen. "What're we looking at right now?" I asked Renato Kane, who was sitting next to me, cleaning a map from earlier in the day. He glanced over. "Uh . . . bad data," he muttered, and went back to removing errors from the maps. These are known as "flyers," when a ping bounces off at an odd angle or records the echo of a previous ping. Kane was doing his best to clean out all the flyers as quickly as possible before new maps rolled in.

As good as a multibeam might be, it cannot overcome the reality of surveying in high seas. The sonar is mounted to the hull of a moving ship on top of a moving ocean. When *Nautilus* crested a wave, the multibeam momentarily lost contact with the surface. Then the ship slammed back down again, but the multibeam was still disconnected because it landed on a wake of bubbles so dense that they blocked the sonar beam. Whenever that happened, the connection between the physical computer (housed elsewhere on the ship) and the screens in the data lab would drop.

After the sonar latched back onto the ocean surface, the water column reappeared on the screen. It went on like that, flickering on and off in time with the waves, as though we were fiddling with the antennae on an old TV. It also meant that we were potentially losing hundreds of soundings in the ocean. "Little lost pings," as one mapper put it poetically. On another screen, those lost soundings smeared an arc under the ship that looked a little like the Joker's infamous messy smile. Kane was working as fast as he could, but the flyers were everywhere.

On the main computer screen there were six separate windows to watch, with two larger windows that showed the ship's current track line and the water column beneath the ship. The water column window looked a bit like a fetal ultrasound. But rather than the grainy black-and-white profile of an unborn baby, it spotlighted the seafloor beneath us in a cone of sound.[14] At the apex of the cone was the ship and then beneath that the surface water in bright royal blue. Farther down came a swarm of turquoise dots: hundreds of millions of plankton, squid, and fish. At night, the swarm rises to feed in the surface waters, and during the day it sinks back down to hide from predators in the dark.

This is the world's largest daily migration by biomass. Scientists call it the deep scattering layer because of the way the sonar ricochets off this wall of life. During World War II, scientists working with the US Navy first spotted it while running sonar experiments and hunting enemy submarines off the coast of California. The layer was so thick with fish and plankton that it looked as though the seafloor was moving up and down throughout the day. The scientists who discovered the layer dubbed it the "false bottom."[15]

The scattering layer is quite beautiful to watch. One mapper on board *Nautilus* told me that she clocked her day around that layer, watching the animals move up and down the ocean, the underwater equivalent of the rising and setting sun.

By its very nature, mapping the seafloor involves introducing more sound into an already noisy ocean. A wide cross section of marine animals shows a range of ill effects from exposure to human-made noise. One experiment exposed scallop larvae to seismic pulses and showed abnormal development, including body deformities, in about half the larvae.[16] Boat noise impairs the embryonic development of sea hares (a type of sea slug) and increases larvae mortality.[17] Pile driving, used in offshore construction, startles squid, which can impair the animal's ability to detect and evade predators.[18] However, the flagship animal for studying the underwater racket is the beaked whale, a highly sensitive deep-diving predator that orchestrates unique diving patterns and calls, even coordinating dives with the other members of the pod to avoid their most feared predator, the killer whale.[19] Midfrequency active sonar, used by the navy for antisubmarine warfare, appears to alarm them. The whales stop foraging and echolocating and swim away from the source of the sound at an abnormally steep angle, which could cause them to suffer from decompression sickness, or "the bends"—similar to the illness scuba divers get when they surface too quickly and develop poisonous nitrogen bubbles in their blood. Since midfrequency naval sonar was introduced in the 1960s, there have been dozens of beaked-whale strandings linked to naval exercises or occurring near navy bases and ships.[20]

"At the moment, a lot of research has gone into understanding the effects of naval sonar and shipping noise on marine mammals," said

Annamaria DeAngelis, a NOAA biologist. These are the two types of noise pollution that have the most obvious impacts on marine mammals. According to DeAngelis, the scientific community tends to regard mapping sonar as low impact because the sonar's range is limited to the water beneath the ship. But the research is still early, and DeAngelis and others are running experiments to test whether that long-held assumption is true. She suggested swapping out survey ships for autonomous underwater vehicles (AUVs), because they travel underwater closer to the seafloor and impact a smaller area.

Back on the main computer tracking *Nautilus*'s sonar, at the very bottom of the water column, came the seafloor: a wobbly red line that revealed that the pings had hit pay dirt. The ways the sonar pings bounce off hard rock and soft mud are different. In the marine science world, this is called backscatter data. Think of it as being like the acoustics in your local bar: a gleaming new microbrewery, covered in glass and steel surfaces, might look good, but it sounds terrible, the hard surfaces ricocheting music and voices into a cacophony. But the softer wood-paneled interiors of your local dive bar are far kinder on the ear. In this example, the hard surfaces of the microbrewery would be rock, while the soft surfaces of the dive bar would be mucky sediment. Backscatter data are a hot commodity for ecologists, who use them to predict where animals live. Deep-sea worms like soft sediment that they churn up while looking for food. Anemones latch onto hard surfaces and stretch their tentacles out into the current, also in search of their next meal.

In smooth sailing, the multibeam creates a rainbow-bright map of the seafloor—a bathymetric map, to use the proper scientific term. A bathymetric map is a hallucinogenic thing to behold. The EM 302 lights up the seafloor with sound, casting a rainbow across mountains and valleys, each color indicating a different depth. The reds and yellows represent the shallower seafloor; then come the greens, the blues, and finally the purples of the deep. This is similar to the color coding found on terrestrial maps, but to the average viewer, bathymetric maps look surreal, sci-fi, almost otherworldly.

I've heard ocean mappers wax poetic over the beauty of a seafloor map, describing a new chart of an Alaskan fjord as though it were a great work of art. They know how much work went into mapping the seafloor, so every granular dimple and pockmark in the bottom is precious. They also usually have a geological background, so their eyes are trained to read the terrain for clues about the distant past and the coming future. They might be able to point to where an ancient river once drained into the ocean or where an island eroded into a seamount. They might be able to warn you about the buckling sides of a trench and a coming tsunami. All of this invisible earth, covered by ocean, is suddenly made visible.

"I think it can be hard for people who see the rainbow-colored maps to get a real image [of the seafloor]," Nicole Raineault told me. "But imagine draining the ocean and being able to see all the mountains, all these Grand Canyon systems that are on the seafloor. It's incredible to see that. And in many cases where we're mapping, we're the first people to see that."

But to be honest, the maps we were making off California that day were not so pretty. The survey looked like a long, ragged technicolor tiger stripe. On the second day of the mapping cruise, Renato Kane decided to shut down the EM 302 altogether, something that was rarely done. "We don't shut it down very much. It's grumpy, like an elderly person. You don't fuck with it," explained Erin Heffron, another mapper on board. By that point, though, we were collecting more bad data than good. Rather than send bad maps of the seafloor to Seabed 2030, the mappers chose to wait for calmer seas.

5.

Later the next day, the weather improved. The swells shrank back into the sea. The wind eased. The captain allowed people to walk outside on the decks again and get some fresh air. I seized the opportunity. As I strolled along the top deck, the sun peeking weakly through clouds, I ran my hand absentmindedly along the deck railing. When I pulled it back, my palm was covered in salt crystals. Over the last twenty-four hours, the gale had brined the ship.

Erin Heffron, the mapper on evening watch that night, was down in the data lab, gearing up to turn the EM 302 back on. "Usually I say a prayer or, you know, make an offering," she joked. We went into a cold temperature-controlled room just off the data lab that looked a little like a server farm with aisles of electronics standing at eye level. That was where a wall of computers, controlling the EM 302, lived. Turning it on was not as easy as flicking a switch, although that was how the ten-minute process began. Heffron reached out and pressed the big black "on" button. And so we began to map.

CHAPTER 3

TO THE BOTTOM OF THE ATLANTIC

1.

Cassie Bongiovanni's new workstation on board DSSV *Pressure Drop* was a simple desk with four computers bolted to the wall. Around her, engineers were fine-tuning the electronics on what could soon become the first submersible to dive the bottom of all five oceans. Biologists were setting up microscopes to pore over never-before-seen deep-sea critters. From that desk, she would start her search for the absolute depths of the planet. On her first day aboard Victor Vescovo's ship, she clung to that desk as if it were a life raft, terrified of screwing up her first job out of university. She was too nervous to leave the dry lab—even to use the washroom. "If I'm not introduced to a space, I assume it's off-limits," she said. Someone eventually showed her where the toilet was. She kept her head down as the drama of the Five Deeps Expedition unfolded around her.

Cassie had arrived in Curaçao the night before she was due to start the job. The arrival had been all a "bit more sketchy than I had antic-ipated," she said. Technically, she belonged to the ship's crew, so she traveled the way ship crew the world over travel: she received a plane ticket but no instructions on where she was staying or how to get to her accommodations once she arrived. After she landed at the airport, security pulled her aside to ask about the extra hard drives Five Deeps had asked her to bring with her. After an intense round of questioning, she was let into the country. She stood on the curb of the arrivals terminal, feeling disoriented and looking out at the dark Caribbean night, whirring with a sound of insects louder than she had heard in her whole life. Then a man emerged from a large SUV in the parking lot. "Are you Cassie from Five Deeps?" he asked. He deposited her at a hotel with instructions to be ready for pickup at 6:00 a.m.

When the sun came up the next morning, she began to get her bearings. Willemstad, the capital city of Curaçao, is an odd clash of Dutch architecture painted in Caribbean Popsicle colors. *Pressure Drop* was docked at a gigantic Willemstad shipyard that services ships heading to the nearby Venezuelan oil fields. After she pulled up to the sprawling shipyard, she wandered amid behemoth oil tankers and container ships, looking for *Pressure Drop*. When she finally found the ship in dry dock, hauled up above the concrete, it was buzzing with workers, people crisscrossing each other on their way to complete a million last-minute tasks before the ship set sail. Who worked for the ship? Who worked for the dock? Cassie walked up the gangway, looking for the captain or whoever else was in charge. A man intercepted her before she set foot on board. "Who are you?" he asked. She introduced herself, and he pointed her onward, deep into the ship.

Built in 1985, the newly renamed *Pressure Drop* was one of a dozen sister ships that had been operated by the US government during the Cold War. Before being converted to peaceful purposes in 2002, it had hunted Russian submarines—another overlap between scientific aims and military ones. The quiet ship had stealth on its side, which made it perfect for sounding the deeps. Victor had paid for a costly retrofit to classify the ship as a civilian-owned vessel. Along the way, he had installed a few creature comforts to make long trips at sea more tolerable. All the cabins came with new flooring, sinks, and flat-screen TVs stocked with a menu of movies. Victor's caffeine of choice is Diet Coke, but he purchased an expensive espresso machine for the ship's crew as a morale booster. An outdoor observation deck above the bridge came to be known as the "sky bar." Beer and wine were set out there each evening as the sun set.[1]

Pressure Drop could sleep up to forty-nine people, most of whom belonged to the Filipino and Eastern European crew. The Scottish officers had worked together in the oil-and-gas industry, including the captain, the straight-talking Stuart Buckle, who had also helmed the ship that had gotten James Cameron his "deepest solo dive" record in 2012.

Another boisterous contingent on board consisted of the engineers and mechanics from Florida's Triton Submarines, who had built Victor's titanium submersible from the bolts up. The Triton crew liked to cluster around a gigantic table right across from Cassie's new work desk. They called that table "mission control," which mostly meant "six to ten guys [who would] all sit around the table and yak at each other about whatever's going on," Cassie said fondly. The scientists managed to get some peace by hiding out in a closet that was just big enough to seat two to three people and their laptops. (It came to be known affectionately as the "science cupboard.")

Even with the upgrades, Heather Stewart said, *Pressure Drop* "was never some superyacht or something. She's an old lady." Alan Jamieson compared working on it to working in an old barn, although he admitted that he had come to love the ship. The expedition crew tucked themselves away in another corner. Cassie, however, sat exposed right at the entrance to the lab, where all the teams collided.[2] "She had a very unfortunate position. I don't know why they put her desk in the middle of the lab," said Jamieson pityingly. "It's like being in the middle of a train station or something." He also felt badly about the bait he stored in a refrigerator right next to her desk. It became a running joke that whenever Jamieson opened the fridge door, the pungent odor of some long-dispatched mackerel hit Cassie square in the nose.

At that early stage of the expedition, the mood on *Pressure Drop* was like in a pressure cooker at sea. There were various gripes among the teams, due mainly to the haphazard way Five Deeps had come together. Triton Submarines had had an early financial stake in Five Deeps and a clear-cut goal that trumped the other teams' priorities: to build the first submersible that could get Victor to the bottom of all five oceans. Triton CEO Patrick Lahey and his team said they would do whatever it took to make that happen. The Scottish ship crew, however, thought the Triton team was far too casual about safety on board,[3] while the captain felt he had been stuck with a subpar ship that Triton had advised Victor to buy. Meanwhile, the scientists were hungry to get their hands on valuable deep-sea samples, but that would come second to Victor's diving the deep. After an off-putting introduction to Five Deeps, Alan Jamieson was still skeptical about the scientific aims of the project. "It was actually quite offensive," he said of an early phone call he had about the expedition. The pitch, according to Jamieson, was "We need a scientist on board, because in a year's time, we want to sell the sub and we want to make it look like it's useful. But don't expect to get any science, and you're never diving in the submersible."

On top of those tensions, Victor invited a documentary crew on board to film the expedition for the Discovery Channel. The film crew elbowed its way into tense situations, asked off-base questions at the exact wrong moments, and generally tried to light up what was already a powder-keg situation for on-screen fireworks. The film crew would soon become the ship's scapegoat, the one group most people on board could agree to dislike. They took over a corner of the lab, on the other side of Cassie, where she listened to them review footage all day.

Cassie, meanwhile, had her hands full with Kongsberg's technicians, who were installing the new EM 124 into the bowels of the ship. As she watched them wire it up, she kept careful notes on what they were doing so that she might have a hope of repairing it later on, if worse came to worst. The technicians would accompany Five Deeps on the first leg of the trip, but after that she would be on her own with the very expensive new sonar EM 124.

A week after Cassie arrived, the sonar now installed, *Pressure Drop* sailed out of Curaçao. It was headed first to San Juan, Puerto Rico, to pick up the remainder of the crew. Then it began a twelve-hour trip out to the Puerto Rico Trench, which lies about 75 miles north of the island.[4] Five Deeps was behind schedule, so the captain throttled the engine to top speed: a grand total of 10 knots (11.5 miles per hour). The ideal speed for mapping the seafloor is a little slower, between 5 and 8 knots (5.75 to 9 miles per hour). That meant Cassie could run and debug the EM 124 only in the off-hours, usually at night, waking the Kongsberg technicians to help her when things went wrong. And things often went wrong.

The marine geologist Heather Stewart was one of the few people on board who understood what Cassie was up against. "This is either going to be awesome, or it's just going to be the biggest disaster ever," she remembered thinking.

After the Mariana Trench's Challenger Deep in the Pacific, the Puerto Rico Trench was the second-deepest deep on the expedition's whole itinerary.[5] At around 5 miles deep, the Puerto Rico Trench is about as deep as Mount Everest is high. Compared to the uncharted depths of the Southern and Indian Oceans, the Puerto Rico Trench had a sizable stash of maps for Cassie to draw on, some dating back to 1876, when the British ship *Challenger* had first discovered the trench. In 1939, the US ship *Milwaukee* had discovered what its crew thought was the Puerto Rico Trench's deep, later naming it the Milwaukee Deep. There was also another candidate known as Brownson Deep. Remarkably, people had even dived the Puerto Rico Trench before, back in 1964, when the French research submersible *Archimède* had visited it.[6]

All the old and new data gathered in the Puerto Rico Trench could help Cassie, but it could hurt her, too. The measurements could be unreliable, culled from shoddy work, made with outdated or poorly calibrated equipment. She also needed two days to run what's called an acceptance test to confirm that the EM 124 was calibrated correctly. "When I told them [the acceptance test] was going to take forty-eight hours . . . jaws on the floor, pissed off," Cassie remembered. A single day on board a research vessel costs upward of $50,000.[7] It was the exact situation Cassie had worried about before joining Five Deeps: Would the older, higher-ranking men on board listen to a twenty-five-year-old female mapper fresh out of grad school? "I made really big arguments to do as much mapping as possible before we even do anything," she said. She stressed to Victor that any deep-diving history he might make in the Puerto Rico Trench, or anywhere else in the world for that matter, could be questioned unless she ran the acceptance test first. She won the argument.

Before the ship reached the deepest parts of the trench, a final series of test dives was done in shallower water to confirm that the submersible was indeed ready to dive. Over the coming days, a comedy of errors ensued. The submersible's hatch leaked water. The electronics were tetchy. If it hadn't looked so dangerous, launching and recovering the submersible would have been an outright farce. That added even more friction to the fight between the Triton team, in charge of deploying the submersible, and the ship's crew, in charge of safety on board. During one deployment, Victor climbed from the launching dinghy onto the floating submersible and became so nauseated, he threw up into the pitching sea. The dive was aborted, but when the seasick Victor tried to climb off the submersible and back into the dinghy, he slipped into the ocean. If a wave had swept through at that moment, the submersible would have crushed him against the dinghy. Luckily, he scrambled out of the water before the next wave passed through. The film crew moved in for a close-up, craning for the best shot as the situation deteriorated.

All the errors and miscalculations mounted into a coup de grâce on the third and final day of test dives, when everything had to go off without a hitch. There was only one more day left in the schedule for Victor's record-setting dive. Patrick Lahey joined Victor on board the submersible *Limiting Factor* as it sank 3,280 feet. The dive went well at first. The hatch had stopped leaking. The electronics were cooperating. When they reached the bottom, the two men peered through the viewports at the strange deep-sea creatures floating past. Victor went to use the sub's manipulator arm. It broke off the submersible and plopped unceremoniously onto the seafloor.

"Patrick, we just lost the arm," Victor said.

"Nooooo," said the horrified Lahey, peering out the viewport. The submersible had come with only a single arm, so there was no way to retrieve the arm once it fell off. Losing the 220-pound arm turned *Limiting Factor* buoyant, and it started an unstoppable ascent to the surface. "I don't know where we go from here, Patrick," the visibly frustrated Victor said to Lahey as they drifted upward.

Losing the manipulator arm was particularly crushing for the scientists on board. Alan Jamieson and Heather Stewart had planned to collect samples with it; the loss threw off all their plans. For Victor, the $350,000 price tag for the part was yet another frustrating sunk cost. "There was a collective feeling on board the vessel that this is never going to happen. Everybody had essentially given up on us at that moment. This thing will work when pigs fly—that's what people were saying," said Lahey. He had more than forty years of experience troubleshooting submersibles, so he was familiar with the trial-and-error approach. But to outsiders, the flaws might signal something more serious: Was this submersible even *safe* to dive all five oceans? And without a working submersible, there was no Five Deeps.

Back on board, Victor called Lahey into a make-or-break meeting with him and Rob McCallum. After the millions of dollars of cost over-runs, the missed deadlines, the damage claim from the worker who fell through the hatch, and now the submersible arm lost on the seafloor, Victor threatened to call off the expedition altogether. Lahey pleaded, "You got to give us a chance. I think that we can fix this and we can get you your deep." Victor eventually relented and gave Lahey another thirty-six hours to repair the submersible. Then Victor holed up in his cabin. The Triton team got to work at breakneck speed on the final repairs. "I'm relentlessly optimistic," Lahey proclaimed. "But I do think that at that moment relentless optimism is what was required. I'm not delusional. I just believed in my team and in their ability to fix this problem."

The extra hours gave Cassie a window to locate the deepest deep in the Puerto Rico Trench. She scoured the bottom, looking for that superlative depth—the one place in the Atlantic Ocean where Victor could not dive any deeper. Typically this is a job done by a team of three or four mappers who swap off day and night shifts, as on *Nautilus*. One person might oversee the multibeam and correct the incoming data; another might analyze the maps, write up reports, and attend logistical meetings. In Puerto Rico, Cassie did it all. She barely slept the entire time at sea.

Only once the ship arrived over a potential deep could she begin to survey. In order to cover as much terrain as possible, *Pressure Drop* had to motor quickly to the most promising spots in a trench that spanned more than 500 miles—about the distance from San Francisco to San Diego. Jamieson and Stewart directed the ship toward a site they be-lieved held the deepest point in the trench, based on an earlier paper they had published. But "it was no deeper than anything else we had mapped at that point," said Cassie. "So is there a deeper spot? How are we supposed to figure this out?"

An ocean trench is a long fissure in the seafloor with a steep drop-off at the rim and a mostly flat bottom.[8] Cassie's job was to find the deepest point of that mostly flat bottom, but even a state-of-the-art multibeam struggles to differentiate between two very similar spots right next to each other. For all the thousands of pings that Cassie rained into the Puerto Rico Trench, each sounding came back with a statistically generated margin of error that was equal to the difference between the trench's deepest and shallowest points. In other words: it was a nearly impossible task. Cassie started to run east-west survey lines across the trench, looking for a downward slope that might lead her to the absolute bottom of the Atlantic. In total, she surveyed nearly 1,544 square miles of the trench—roughly the size of Rhode Island. By morning, she found the deepest point tucked between two earlier survey lines: 8,736 meters (28,661 feet) deep, give or take 5 meters (16 feet). "That one was really hard to find," she said with a tired exhale.

Meanwhile, the Triton team discovered that losing the submersible arm was actually a windfall. "The arm falling off actually created the solution," Lahey explained. The mechanical arm required more than thirty conductors to function, but now those conductors could be rewired to address other finicky problems inside the submersible. "Thirty-six hours later, Victor climbed into the sub and he didn't have a single alarm on his alarm panel. Whereas on the previous dive, it was lit up like a Christmas tree. Every system that hadn't been working correctly was now working correctly, except for the arm."

With the Puerto Rico deep pinpointed and a submersible that appeared to be in good working order, Victor announced that he would dive the trench after all. (He later insisted that he had never planned to call off the expedition. "I did that to provoke them," he said. "It was just a motivational tool.") On the absolute last possible day to dive, the skies cleared, the winds calmed, and everything looked perfect for reaching the deepest point in the Atlantic Ocean.

2.

Long before Five Deeps became a real expedition, with a ship, a submersible, and a whole team behind it, Victor had always known he wanted to dive the deeps alone. For someone who is very much in the public sphere, who has summited mountains and commandeered corporate boards, he is happiest on his own. He's never married or had kids, but he does have a troop of small black barge dogs known as Schipperkes that he treats like kin. Going by himself also meant that Victor's world record would be his alone. Patrick Lahey, an extrovert's extrovert who thrives in the middle of chaos and crowds, had never liked the idea.

"I don't really understand why people want to do things alone. I think most human experiences are made richer when you share them with others, but [Victor] wanted to do these dives on his own. So now, not only do we have to train him to be a pilot and to pilot these dives on his own, we had to design systems in the vehicle that would address the eventuality of his incapacity." In other words: What would the team do if Victor became unreachable thousands of feet beneath the surface? The farther down he dived, the harder it would become for the ship to rescue him. If the absolute worst-case scenario unfolded, how would the crew retrieve his body?

The descent into the Puerto Rico Trench would be Victor's first time piloting the submersible alone, and he would dive deeper than he had ever gone before. After *Limiting Factor* sank beneath the surface, the team retreated back to the dry lab to watch and wait the three hours until Victor touched bottom. Every fifteen minutes, he was supposed to call the team for a communications check. His short, clipped check-ins came with a delay as the sound passed through the widening gulf of water between the ship and the submersible. Each time, he read out his current depth and heading and reported that all life support systems on board were good. About halfway through the dive, when Victor was about 4 miles beneath the surface, he missed a check-in. Lahey radioed him. "Victor, can you read me?" he asked urgently into his headset. Nothing. Ten minutes passed as Lahey hailed Victor again and again and listened to a wash of static. The tension in the dry lab grew and grew, everyone waiting in silence as unspoken questions hovered in the air. At that depth, there would be no way to rescue Victor if something went wrong. Had all those tiny glitches with the submersible been warning signs of a more serious breakdown to come? Had they just lost Victor at the bottom of the Atlantic?

Lahey put his head in his hands. "Oh, my god." A full twenty-five minutes had passed since Victor's last communications check. Lahey tried to hail Victor on the headset again. "Victor, can you read me? Can you read me?" Suddenly Victor's voice cut through the static. He had been calling, he said, but he would speak up next time. Everyone in the dry lab exhaled a collective sigh of relief. Lahey, who looked as though he had been on the verge of a heart attack, regained his composure.

Victor had a far different experience, describing his journey to one of the most extreme limits of Earth as quiet, peaceful, almost serene. His submersible sank through the ocean at a rate of 2.5 feet per second.[9] He watched the instruments on the submersible's dashboard flicker and the darkness pass by outside the viewport. "In general, it's quiet," he said.

The ocean is divided into three zones—the Sunlight Zone, the Twilight Zone, and Deep Sea—based on how much light filters down from above. On the first part of his journey, Victor passed through the Sunlight (or Epipelagic) Zone, the surface ocean named for the sunlight filtering down from above and the plant and animal life photosynthesizing off the sun's rays. This is the ocean that most of us are familiar with. It's where we catch our seafood, know the most species, and can explore within limits. At around 200 meters (660 feet) deep, Victor entered the Twilight (or Mesopelagic) Zone. This is the deepest humans can descend without some kind of protective craft. In 2014, an Egyptian special forces officer visited the Twilight Zone, earning the world record for deepest scuba dive at just over 300 meters (1,000 feet) deep. It took him fifteen minutes to descend and thirteen hours to ascend, taking time for decompression stops lest his lungs explode and his blood boil due to the changing water pressure.

The dusk of the Twilight Zone is well suited to the bioluminescent marine life that flickered and flashed around Victor's submersible as it passed by. Whereas most submarines have an oblong shape that allows them to explore horizontally in the three-dimensional space of the ocean, Victor's slim submersible was designed to move quickly up and down the water column. The bright white submersible looked a little like a gigantic molar, tapered at the top with a bulb at the bottom to hold two passengers. Inside *Limiting Factor*, the temperature dropped rapidly, signaling that Victor had reached the thermocline, the boundary between the warm, well-mixed waters above and the cold, sluggish waters of the deep.

At the 1,000-meter mark (3,280 feet) he entered the third and final zone: the Deep Sea. All light from above disappeared. The pressure pushing down on *Limiting Factor*'s titanium hull now reached more than 1,400 pounds per square inch. In the navy submarine world, this is known as the "crush depth": any vessel that is intended to go deeper requires fancier engineering and testing.[10]

The top two layers had taken Victor only twenty minutes to pass through. Now he had another 7,000 meters (22,965 feet) more to fall before touching bottom. The Deep Sea encompasses 75 percent of the ocean's depth, which is why deep-sea specialists such as Alan Jamieson find this nomenclature far too vague. Instead, experts divide the Deep Sea into three more zones, called the Midnight (or Bathyal) Zone, the Abyssal Zone (or Abyss), and finally the Hadal Zone.

The first layer, the Midnight Zone, extends from 1,000 to 3,000 meters (3,280 to 9,842 feet) in depth. No sunlight reaches the Midnight Zone. Food is scarce, the pressure is high, and the temperature is a chilly but stable 40 degrees Fahrenheit. The Midnight Zone is a difficult place in which to survive; its most famous denizen is definitely the anglerfish, with its set of fearsome backward-pointing teeth and its dangling light to lure prey to it. At that point, Victor was diving along what's known as the continental slope, where the seafloor rapidly drops off from the landmass.

An hour into the dive, Victor arrived in the Abyss, which reaches 6,000 meters (19,685 feet) deep. In 97 percent of the ocean, the Abyss is as deep as it gets. Its name stems from a time not too long ago when people believed that there was no bottom to the ocean, just an abyss, an endless, terrifying chasm. Much of the Abyss is a flat plain of soft, silty sediment, interrupted here and there by undulating valleys, active volcanoes, and flat-topped seamounts known as guyots. But Victor had still farther to fall, dropping past the Abyss and into the Hadal Zone, named after Hades, the Greek god of the underworld.

Deep-sea trenches are rare. Less than 3 percent of the global sea-floor falls within the Hadal Zone. The name's allusion to the under-world is fitting. Geologically, the Hadal Zone is where the seafloor goes to die. With a few important exceptions, ocean trenches are located at seismically active subduction zones, where the seafloor is sucked down into the Earth's molten core and recycled. The trenches also cause death and destruction, shuddering with earthquakes that can trigger tsunamis. In the Puerto Rico Trench, the North American Plate is sliding ever so slowly under and up against the Caribbean Plate. The subduction zone between the two plates creates the deep trench that looks like a knifelike slice in the seafloor. Ocean trenches line the Pacific, and earthquakes are common in the Ring of Fire, which runs along the west coast of the Americas and the east coast of Asia. Most of them are small, such as the five thousand–some earthquakes that strike in the Mariana Trench each year. Every so often, a big one occurs, such as the 1918 earthquake in the Puerto Rico Trench that sent a towering wave ashore and killed more than a hundred people in Puerto Rico.[11] They are also devastating for the marine life in the trench, having the brutal effect of hitting the refresh button on an entire ecosystem.

As Victor closed in on the bottom of the Puerto Rico Trench, he noticed light coming in through his downward portholes. He looked outside and saw that his submersible's lights were bouncing off the bottom, even though he couldn't see it yet. "Things get busier when you get within the last five hundred feet. Then you better start adjusting the ballast and also starting to triangulate where you are," he later said. He watched through the porthole anxiously as a muddy moonscape loomed up beneath him. As the seafloor raced closer and closer, he prepared for landing. It was a soft one—so soft, in fact, that the sub-mersible sank deep into the muck, covering the headlights and plung-ing him into darkness.

"Surface, surface!" he shouted over the radio. On the bottom, he re-ported, all life systems were good. Miles above him, the crew in the dry lab exploded in cheers. There were high fives, hugs, and handshakes all around. The relief, remembered Heather Stewart, felt like a wave breaking over the room.

Down below, as the muck settled, Victor looked around at where he had arrived. "It's like I'm on another planet," he said to himself.[12] The flat lunar landscape; the dark, heavy ocean hanging above like a night sky; the submersible hovering over the seafloor a little like a spaceship—it all felt strangely reminiscent of footage beamed back from outer space. The journey there had even looked a little like space travel, too. As Victor had sunk through the ocean, moving at roughly 150 feet per minute, plankton had raced past the viewport, looking like stars moving at warp speed from a rocket ship blasting into space.

Victor began to roam around a place that no one had ever visited before. Using the joystick, he started to move over the seafloor, the submersible's high beams illuminating one patch of ground at a time like the flashlight beam of an intruder moving around a dark house. A mountain summit is the classic endpoint of adventure. At the very top, where the explorer can climb no higher, she is rewarded with a view over the land below: a chance to survey how far she has come, physically and psychologically, perhaps as the sun breaks through the clouds, adding another beautiful dash to the scene. The deep sea can't offer that sort of vantage point. There are no spectacular vistas at that depth; even the nearest geographical features are shrouded in darkness. Victor could see about a hundred feet in front of him, so he explored with the submersible's nose to the ground, shuffling along the seafloor inch by inch. Later, it occurred to him that he had had a similar view at the top of the world. "Some people go to Everest on a clear day, they can see for fifty miles," he said, but that hadn't happened for Victor, who had summited the peak during a storm. He could barely see two hundred feet in front of him through the swirling snow.

Some speculate that the ocean's dark obscurity is what makes exploring space more popular and accessible. "With a telescope and a clear night, any of us can get a reasonable idea of what the near side of the moon looks like. . . . Try doing the same for deep seabed," writes the marine biologist Helen Scales.[13] These preferences are baked into the way we express ourselves. When we feel down or in over our heads or we're scared of what lies beneath us, the negative connotations are clear. The reverse is true for when our spirits lift, when we feel as though we're on top of the world or back on solid ground. Up is better than down, light is better than dark, and height is better than depth. These are all situations that play to our species' strong points. We naturally feel more in control when we can see something, and we feel that we understand it better, too. The ocean is the quintessential dark place, while space is the ultimate escape in movies and video games. Even when humans have to battle hostile aliens, outer space is still a mostly fun backdrop. The ocean, meanwhile, is more often featured as a sort of hellscape in horror movies and bleak documentaries about overfishing and pollution.

At a TED Talk in 2008, Robert Ballard, the founder of the Ocean Exploration Trust and owner of *Nautilus*, homed in on these issues: "Why are we ignoring the oceans?" he asked. "Why are we looking up? Is it because it's Heaven and Hell is down here? Is it a cultural issue? Why are people afraid of the ocean? Or do they assume the ocean is a dark, gloomy place that has nothing to offer?"[14]

Up ahead, Victor spotted a round black object sunk in the sediment of the Puerto Rico Trench. He piloted the submersible over to investigate, and his lights illuminated what looked like a cracked oil drum. Someone had been here before, or at least his trash had. Victor wrinkled his nose in disgust, like a hiker who finds his mountain peak sullied by discarded candy wrappers and beer cans. Here is yet another reason we're drawn to outer space: there are no humans there. Apart from the space junk orbiting Earth, the universe is devoid of other people and all their baggage (and garbage). We can imagine entire untouched extraterrestrial landscapes. The ocean, on the other hand, has long suffered as humanity's dumping ground.

Victor insisted that none of that—the limited outlook, the unimpressive seafloor, the trash—had dampened his enthusiasm for reaching the deepest place in the Atlantic. "I've traveled all over the world during my life," he said. "I've gone from desolate deserts in the high mountains to jungles and the Serengeti, and they're each special in their own way. Whether it's stark or whether it's arid and desolate, they all have their own beauty. For me, it's just like looking at a different color." The thing that really grabbed him was that no one had ever been in that exact place before. "Just the act of discovery was fascinating for me."

After Victor had spent about forty-five minutes touring the bottom of the Atlantic, Lahey raised him on the underwater comms and recommended ending the dive. Victor agreed and flipped a switch on board, jettisoning weights that kicked off his ascent. Ironically, the deep's first visitor had to cast off his own waste to return from it. It took two and half hours to return to the human world. That evening, just after 6:00 p.m., the submersible floated up from the depths.

"There was this noticeable change in everybody's attitude" after Victor climbed out of the submersible, remembered Lahey. "Suddenly all the people who thought this was never going to happen thought, 'You know what? This might actually happen.' Including Victor."

3.

Back on board, Victor brought Cassie some good news: the depth readers on *Limiting Factor* had reached a maximum depth of 27,480.5 feet—only a half foot off her estimates. "Cassie was more surprised than I was," he remembered. "She knows the limitations of the system. I thought that was kind of normal, like 'Why *hadn't* we expected that?'"

As a longtime mountaineer, Victor had a natural affinity for maps. He often stopped by Cassie's desk to watch her work. There was something about the seafloor maps he found particularly compelling. The ones he had used to scale mountains on land looked similar to the ones he was now using to reach the ocean's depths.

While she worked, Cassie casually mentioned another effort to Victor: Seabed 2030 and its quest to map the entire seafloor by the close of the following decade. Victor had never heard of it, but back then, very few people had. Even fewer had heard of the group behind Seabed 2030, the General Bathymetric Chart of the Oceans (GEBCO).

Founded in 1903 by Prince Albert I of Monaco, GEBCO has tried for more than a century to bring all the disparate maps of the world ocean together into one giant supermap. A devoted sailor and early oceanographer, Prince Albert was something of a precursor to Jacques Cousteau and other ocean explorers. He outfitted his luxury steamer yacht, *Hirondelle*, with a laboratory, a sounding machine, and dragnets to run experiments.[15] In 1880, he discovered the spinning gyre of the North Atlantic by tossing bottles and barrels into the East Atlantic and discovering that all of them returned from the south.[16] At the time, the first major geography conferences were beginning to be held, and the study of the ocean was only gradually distinguishing itself as a separate field. Meanwhile, calls were growing for a formal organization to take charge of all the seafloor maps created during the efforts to lay the first telegraph cable across the North Atlantic. There was no money behind such an effort until Prince Albert stepped in to finance and publish the first edition of the seafloor map himself, designated as the Carte générale bathymétrique des oceans (General Bathymetric Chart of the Oceans). In 1905, GEBCO, as the organization later came to be known, published the first map of the seafloor. It had fewer than twenty thousand depth measurements across all five oceans.[17]

Over the coming years, subsequent editions took longer and longer to publish. The third edition, interrupted by two world wars, took nearly twenty years to finish. The Cold War and the rise of antisubmarine warfare should have been a boon for ocean mapping, but much of the newly acquired data was classified. GEBCO's mission to share maps was running headlong into a historical trend to hide maps. "Navigators were reluctant to put down on paper what they had discovered, with the result that printed maps and charts were always scarce and there was often a lag of from two to twenty years between the date of a new discovery and the time it became incorporated on the map," wrote the map historian Lloyd Brown about the Casa de Contratación de las Indias in Seville, the trade organization that included the world's oldest hydrographic office, including Spain's colonial-era mapmakers, who attempted to create a master map of the New World.[18] That was what GEBCO was up against: centuries of secrecy surrounding seafloor maps. A pattern soon established itself: by the time each new GEBCO edition was published, it was almost instantly out of date and work had to begin on the next edition.

By the mid-1960s, GEBCO was embroiled in a fight among its own members. The growing ranks of academic mappers started to challenge the government and military mappers who had long overseen the production of GEBCO maps. The geophysicists, oceanographers, seismologists, and volcanologists saw GEBCO's narrow focus on safe navigation as outdated. There were exciting new discoveries, the most important of which was the advent of plate tectonics in the late 1960s, which placed the seafloor at the forefront of new geological studies. Fewer and fewer scientists purchased the GEBCO charts, and, faced with dwindling demand and plummeting sales, the organization was threatened with dissolution.[19] GEBCO recovered by incorporating more academic mappers into its ranks, but as it limped into the twentieth century, its noble dream to produce a complete seafloor map was still frustratingly out of reach.

In 2003, the GEBCO mappers gathered to celebrate a century of charting the seafloor. Robin Falconer, a mapper from New Zealand who joined the group in the 1970s and served on various GEBCO committees over the years, remembered a warm anniversary party followed by an unsettling wake-up call: "Immediately after the anniversary, we had an annual meeting and we were sitting in a room and we looked around at each other and said, 'We're all old. Where's the next generation?'"

Falconer was in his fifties at the time and the youngest person in a room of about a dozen mappers. Many of the leading mappers were aging out of the field and retiring. Major scientific funders no longer considered ocean mapping a cutting-edge pursuit. Without fresh funding for new maps and new mappers, how would GEBCO train the next generation?

GEBCO needed to replenish its ranks, but it also needed to diversify. The earlier generation had consisted mostly of white male mappers from developed countries, and the maps they had made reflected that. "I don't want to criticize [GEBCO] too much, but back in the late eighties and early nineties, they were really focused on Europe, like 'Let's go over to Europe, where we have *perfect* coverage,'" said David Sandwell at Scripps. This isn't necessarily the fault of the mappers; it's more the fault of the governments and institutions that employ them. Wealthy countries from the global North have bigger budgets, and they will almost always prioritize surveying their own territory over surveying international waters.

Meanwhile, the charts of poor nations tend to be spotty. Where charts do exist in the developing world, they often date back to the colonial era, when French ships surveyed the waters off Tahiti or British ships surveyed their country's whaling stations near Antarctica. Most modern surveys in the global South are undertaken by scientific research or the military. And the military doesn't have much incentive to share maps with an organization such as GEBCO, whose mission is to create the first complete and publicly available map of the world seafloor.

Not long after the GEBCO centennial in 2003, an executive in Tokyo received a fax inviting him to a gathering in London, England. Mitsuyuki Unno worked in the Ocean Affairs Division at the Nippon Foundation, Japan's largest philanthropic organization, which funds global programs such as eliminating leprosy and creating agricultural initiatives in Africa. According to GEBCO member John Hall, another member from Japan had urged the mappers to reach out to the Nippon Foundation for funding—and GEBCO finally took up the suggestion.

Unno flew to London, where he was shown into a grand room and took a seat on a wooden thronelike chair with lion heads carved into the armrests. In front of him sat ten ocean mappers, almost all of them white men of advancing age. Over the next four hours, the mappers laid out their case to Unno for why the world's seafloor should be mapped: It could solve very local problems, such as building wind farms off the US continental shelf and managing scallop populations in Greenland. It could address big-picture problems, such as understanding the warming ocean currents created by climate change, how and why coastal areas flood, how a tsunami will roar up on land. A complete map of the global seafloor would not solve all the myriad problems facing the planet, but it could help us prepare for the life-and-death choices we will soon face. Heading into an unstable future without understanding the seafloor is like diving into a murky pool without knowing how deep it is.

Unno could barely understand all the technical verbiage the mappers were throwing around. He was in his thirties at the time, and his background was in international development, but he left the meeting impressed by their passion. "They reached a point where they were starting to feel their age," he recalled. "For the very first time, they recognized that they had no one to pass down all this work. I think they were in a panic." The Nippon Foundation later agreed to fund a GEBCO training program at the University of New Hampshire's Center for Coastal and Ocean Mapping.

It was only later, John Hall recalled, that he learned about the Nippon Foundation's founder, Ryōichi Sasakawa, and his place within Japanese history. In 1945, Allied forces arrested and charged Sasakawa as a Class A war criminal, defined as crimes against the peace, for being "one of the worst offenders, outside the military in developing in Japan a policy of totalitarianism and aggression," as one prosecutor put it at the time. There are wartime pictures of Sasakawa posing with Benito Mussolini, whom he admired and emulated, ordering his private militia to wear black shirts just as the Italian Fascists did.[20] After a change of policy in Washington, however, Sasakawa was released three years later, and his wartime activities were never aired at trial.[21] In the postwar years, Sasakawa thrived by building a gambling empire for motorboat racing and operated as a sort of kingmaker in the surprisingly intertwined worlds of the Japanese mafia, known as the *yakuza*, and right-wing political parties.[22] In 1980, he founded what was known unofficially as the Sasakawa Foundation, handing out generous grants to charities and universities around the world in what seemed to many like an obvious attempt to burnish his legacy.[23] "Despite all his do-gooding and donating, Sasakawa remained a gambling czar with strong ties to the extreme right in Japan, and with less overt but definite ties to the yakuza," wrote the investigative journalists David Kaplan and Alec Dubro.[24]

The Sasakawa Foundation later became the Nippon Foundation after Ryōichi Sasakawa died in 1995 and the left-leaning Japanese government at the time pressured the foundation to change its name.[25] The new name, Nippon, has a nationalistic connotation within Japan, perhaps most directly in the form of the Nippon Kaigi, a powerful ultra-right-wing lobby group that echoes some of the same nationalistic sentiments found in Donald Trump's MAGA movement.[26] The Nippon Foundation is a charitable organization that funds meaningful projects, such as Seabed 2030, but it has also tried to silence historians who raise Ryōichi Sasakawa's wartime record or his ties to organized crime and far-right politics in Japan.[27]

"[The Nippon Foundation's] favorite targets are people who don't know anything about Japan. They love to give money to people who work on nature, on arts, on the environment," the historian Karoline Postel-Vinay told me. In 2008, the French arm of the Nippon Foundation sued Postel-Vinay for libel after she started a petition urging the French Foreign Ministry to withdraw support from a Nippon Foundation–sponsored event.[28] Postel-Vinay won the lawsuit. "In the decade since the court ruled in [Postel-Vinay's] favor, the Sasakawa [Nippon] Foundation has lost influence in France as academics and universities cut ties, and it has been given the cold shoulder by the government," wrote Jeff Kingston, the director of Asian studies at Tokyo's Temple University Japan.[29] Although some universities and groups have turned down money from the Nippon Foundation, many more have accepted,[30] including the University of New Hampshire, where Cassie earned her graduate degree.

Over the last twenty years, the Nippon-GEBCO program at UNH has trained more than a hundred ocean mappers from around the world. The students usually have a professional background in a related field, such as a maritime cartographer from Latvia, a meteorologist with the Royal Malaysian Navy, and a land surveyor from Kenya. However, the Nippon-GEBCO program instructs them in the very specific field of surveying the seafloor. Most ocean mappers I spoke to seemed unaware of the Nippon Foundation's past, although one observed that the GEBCO program had never trained a mapper from China, the world's second-most populous country and Japan's historic adversary. John Hall, the only GEBCO member who directly raised the criminal charges against Sasakawa, also felt his past was ancient history. "I suspect that you can do little good 81 years after the fact dwelling on that aspect," he wrote me. "GEBCO's purpose is getting the damn ocean mapped at a respectable resolution. Seabed 2030 is the first effort that, with present capabilities, stands a good chance."

After finishing the yearlong course, the GEBCO trainees head out on an apprenticeship. Afterward, they either return home to their original job or find work abroad through the GEBCO alumni network, which is now scattered across more than forty countries. The trainees become unofficial ambassadors for the larger GEBCO mission. Back in their home countries, usually working in military or government mapping departments, GEBCO relies on this network to share critical on-the-ground knowledge about how to feed maps from their respective countries into the larger world map.

In 2017, the Nippon Foundation gave GEBCO another $18 million to launch Seabed 2030. That was the same goal that Prince Albert I had set out to achieve more than a century before, but the time seemed ripe for a moon-shot moment in ocean mapping. Later that year, the GEBCO alumni team won the $4 million Shell Ocean Discovery XPRIZE for the best autonomous ocean-mapping technology.

In 2019, Seabed 2030 announced that its latest map, known as the "global grid," now covered 15 percent of the ocean floor at the desired resolution. The last time GEBCO had published its global grid, in 2014, the map had covered just 6.4 percent of the entire ocean at an equal resolution. Between 2017 and 2019, Seabed 2030 had doubled the amount of mapped terrain and increased the detail. At meetings, GEBCO members began to note wryly that their baby, Seabed 2030, had surpassed its parent. In two years, Seabed 2030 had attracted more attention to the uncharted oceans than GEBCO had in more than a century.

4.

As *Pressure Drop* headed back to land and Cassie finished explaining the push behind Seabed 2030 to Victor, he immediately agreed to help. Until then, he had had no plans for the maps collected on board Five Deeps. As much as he might like looking at the maps, they were mostly a means to an end: to become the first person to dive all five deepest points in the ocean.

Now that Victor had ticked off the first deep on the list, the team was riding high on his success. Alan Jamieson still couldn't believe that the dive had happened at all. "It was a mess until the last day," he said. "I don't know how on earth [Triton] did it, but they did it." After months of uncertainty, the expedition found itself on solid footing at last.

Almost as soon as everyone stepped off the ship, a threat to that newfound camaraderie reared up. After a few celebratory social media posts about the dive by Triton Submarines and EYOS, Rob McCallum's expedition company, the film production company sent an angry email threatening to sue the two companies if they posted about the expedition prior to the documentary's premiere on the Discovery Channel. That set off a power struggle among the various teams, who already resented the production company. Eventually Victor managed to get everyone to obey the production company's demands, but not without some bruised egos along the way. Unable to share their own pictures and videos online, Five Deeps now had to rely on reporters to tell their story to the wider world.

Back in the United States, Cassie dug into the first articles published about Five Deeps. The stories spoke of a multimillionaire adventurer chasing a world record to the bottom of the ocean, about new species discovered in the deep and new terrain glimpsed through a submersible porthole. The stories rarely mentioned the shy, sleep-deprived mapper calmly guiding the expedition to the deepest point on the map.

Unfortunately, the optics of mapping the ocean floor do not look all that thrilling. An ocean mapper is typically found seated at a desk in front of a bank of computers buried deep in the bowels of a ship. Ocean mappers, too, admit that running survey lines back and forth over the seafloor can be a bit dull. They call it "mowing the lawn." Hydrography, the scientific field of surveying bodies of water, as well as its equally unsung offspring, bathymetry,* do not look like the rocket liftoff moments that attract massive government funding and effusive public interest. It dawned on Cassie that barely anyone understood her

---

* I like to distinguish between hydrography and bathymetry by using the mariner proverb "A ship can be a boat, but a boat cannot be a ship." Similarly, hydrography can be bathymetry, but bathymetry cannot be hydrography. Bathymetry is the measurement of the seafloor's depth and has come to mean the survey of the overall topography of the seafloor. Hydrography includes bathymetry and many other measurements, such as the observing vessel's vertical and horizontal positioning, the ocean's chemistry, tides, and currents. In general, the maritime world prefers to use hydrographic measurements over bathymetric ones.

role in the expedition. "I didn't realize how much I had hurt myself until then," she said.

The media's oversight fit with GEBCO's obscure history and its century-long struggle to rally the world around charting Earth's final frontier. One story in particular irked Cassie. The piece was meant to headline the mapping efforts on board the ship, and it included a photo. "I'm next to Heather [Stewart], teaching her how to use the [mapping] software. It's eight or nine in the morning. I have been up all night. I'm haggard," Cassie said. "And they cut me out of it and used that picture to advertise mapping on board. I was like, 'You've got to be kidding me!'"

Before the upcoming dive in the Southern Ocean, Victor made good on his promise to donate all of the Five Deeps maps to Seabed 2030. In January 2019, he signed an official agreement so that all the mapping Cassie did on board would become part of GEBCO's growing map. In return, the GEBCO training program at the University of New Hampshire would send a recently graduated mapper to help her on each remaining leg of the journey. Cassie went through the list of candidates and picked a name she recognized from her department back in New Hampshire: Aileen Bohan.

A jolly young Irishwoman, Bohan had come to ocean mapping, oddly enough, through a fascination with outer space. During her undergraduate years at Trinity College Dublin, she had focused first on astrophysics because of her intense interest in the composition of the solar system. She had realized that most of her questions about the composition of comets were actually resolved through geology. After she had gone on a field trip aboard an Irish survey vessel while she was an undergrad, she had switched majors to geology for her master's. Another research vessel trip during her master's sealed the deal and had turned her into an ocean mapper, something she also chalks up to her love of the sea. "In Ireland, if you drive anywhere, you're going to reach the ocean," she said. "We all have this happy love of the ocean."

With an extra pair of hands on board, Cassie would get a reprieve from her round-the-clock mapping duties. She would need the rest on the five-week cruise to the Southern Ocean, swirling around the frozen continent of Antarctica. It is the most rowdy, most remote, and most logistically challenging of all the five world oceans. It could also be the most rewarding for two young ocean mappers eager to make their mark.

While the majority of the crew flew home to rest up over the winter holidays, *Pressure Drop* beat a slow transit south along the east coast of South America. In January 2019, the ship pulled into Montevideo, Uruguay, and the teams regrouped there for the next leg of the journey. The mission was bound for a string of smoking volcanic islets curled like a crescent moon off the southern coast of Argentina: the South Sandwich Islands, uninhabited by humans but home to some of the largest penguin colonies in the world.

To the east of the South Sandwich Islands, the seafloor drops off into the South Sandwich Trench, created by the collision between two tectonic plates. In 1927, a German surveying ship, *Meteor*, discovered what was still believed to be the deepest part of the trench, the Meteor Deep. But because roughly 80 percent of the Southern Ocean is not yet surveyed, even at the very poor half-kilometer resolution, Cassie and Bohan stood a good chance of overturning our basic knowledge about the Southern Ocean.[31]

During the age of sail, merchant ships used to regularly pass through the southern latitudes, using the strong westerly wind at the 40th parallel south, known as the Roaring Forties, to shave days off the trip between continents. Few vessels pass through there anymore, as diesel-powered ships have largely made the wind-powered latitudes obsolete. The crews of the ships that do travel in the area rarely want to tell the world about their trips. Japanese whaling ships go there to hunt, as it's as far away as they can get from international bans. Sometimes Greenpeace activists tail the whalers, swerving as closely as they dare to film the bloodied work decks where the whales are butchered. Scientists pass through on their way to research stations in Antarctica. So, too, do solo sailors taking part in the Vendée Globe, the nonstop round-the-world sailing competition that regularly claims lives. The rest of the time, the Southern Ocean is a lonely place. The farther south you sail, the rougher and faster the latitudes become: the Roaring Forties become the Furious Fifties become the Screaming Sixties,[32] and that was right where Cassie Bongiovanni and the rest of the team were headed.

CHAPTER 4

MARIE THARP AND THE MAP THAT CHANGED THE WORLD

Of what use is a bottom if it is out of sight, if it is two to three miles from the surface, and you are to be drowned so long before you get to it, though it were made of the same stuff with your native soul?

—Henry David Thoreau, *Cape Cod*, 1865[1]

1.

Not long ago, I found a print of one of the most famous maps of the seafloor ever published, selling on Etsy for just $15. Published by *National Geographic* in 1967, the map appeared the same year that a new scientific theory upended our understanding of the planet. By the middle of the last century, ocean mappers had charted enough of the seafloor (although not very much, considering its size) to piece together the first modern glimpse of the seafloor—and the map moved the earth beneath our feet. Until then, most people had thought of the seafloor as being flat, boring, dead earth. The discovery of plate tectonics ushered in a new era in which the most fascinating geology was seen to be in the ocean, hidden by miles of water.

The map I found on Etsy was based on the work of Marie Tharp, one of the first women to work in ocean mapping. Born in 1920, Marie might have gone even further if she had arrived on the scene even twenty years later. But in other ways, she was right on time, drafting the first map that introduced millions of people to the seafloor's diverse terrain: the coastal continental shelves, the deep-sea plains that cover much of the global seafloor, the craggy peaks and troughs of an underwater mountain range that runs a ragged path 40,000 miles around the Earth.

"This was not just an illustration of the ocean floor, but a preliminary explanation of a new geologic theory," wrote the map historian Susan Schulten of Tharp's maps.[2] "One of the most remarkable achievements in modern cartography," proclaimed John Noble Wilford in his classic text *The Mapmakers*, "[a] graphic summary of more than a century's worth of oceanographic effort."[3] Most of the accolades came after Tharp's death in 2006. Although she's now arguably the world's most famous ocean mapper, in her prime in the 1950s and '60s, she was a complicated figure: a divorcée who hid her past, a sharp intellectual who yearned for a job that meant *something*. She struggled to fit in and made compromises along the way, all to draw the maps she loved. Then, at the height of her career, she suffered a devastating loss and found herself professionally sidelined and her contributions diminished.[4] After she died, the *New York Times* and *Los Angeles Times* published obituaries that included discussions of her work. Several children's books, multiple articles, and a full-length biography now recount her adventures, along with a *National Geographic* documentary and a punk rock tribute band. She's also become an icon for the third and fourth generations of female marine scientists. I met a woman who planned to sail to Greenland and survey its melting fjords; she had named her sailboat *Marie*.

Any book about plate tectonics or the seafloor today is incomplete without a nod to Tharp and her close collaborator, Bruce Heezen. Attend a few conferences on Seabed 2030 or read its promotional materials, and you'll hear her name dropped repeatedly. Over time I came to understand why Seabed 2030, and the ocean-mapping community at large, invokes Tharp so often: they are waiting for another Marie Tharp, someone who will collect all the disparate ocean soundings and draw a map that portrays the planet in all its wonderful, complex entirety. In other words, Marie showed us why mapping the seafloor *matters*. I ordered a copy of the map.

2.

The perennial new girl, Marie Tharp was always the outsider when she was growing up. She was an only child born to a schoolteacher mother and a soil surveyor father; her family moved all over the country for her father's job with the US Department of Agriculture. "We were constantly on the move," she wrote, "with Papa working in the southern states during the winter and the northern states in the summer. By the time I finished high school I had attended nearly two dozen schools and I had seen a lot of different landscapes."[5]

The family lived in small towns and big cities across the North and South, including places like Orville, Texas, and Selma, Alabama. Every four years they decamped back to Washington, DC, where the government surveyors gathered to see new measurements made into maps. When she was living in the South, her northern accent marked her as a Yank. She learned to play with the other ostracized kids, including the only Jewish girl in her school and the Black son of a janitor. She turned to books to pass the time, picking up the science magazines lying around the house. On Saturdays, her father took her to survey sites in his work truck. While he mapped and measured, she made mud pies and hunted for skeletons to add to her father's growing vertebrae collection. Once she took a James Fenimore Cooper book along with her. "Why are you reading that when there's all this nature to look at and read?" her father scolded.[6]

She never expected to follow in her father's footsteps. The job options for women at the time ruled it out. Secretary, nurse, and teacher: those were women's jobs, except that Marie couldn't type, couldn't handle the sight of blood, and hated teaching.[7] She drifted through her twenties, into a marriage, a divorce, and degrees in English, music, math, and geology. After earning a master's in geology, she wound up drafting maps for an oil company and then working for the US Geological Survey. In the late 1940s, she walked into Columbia University with the vague idea of finding a job and stumbled into an interview with the geophysicist W. Maurice "Doc" Ewing, who had pioneered the use of sonar with the US Navy during World War II. Doc was even better known for having codiscovered sound channels: ocean layers, created by shifting salinity and temperature gradients, that funnel underwater sound across extraordinary distances from one end of an ocean to another.[8]

"He asked me the standard questions about my background, and he seemed to become more amazed as he heard about my assorted degrees and the order in which I had taken them. His courtly Texas manner could scarcely hide his bewilderment, and finally he blurted out, 'Can you draft?'" she remembered. "And it was ridiculous, he just couldn't understand. He was so bum-fuzzled at this funny collection of courses and in my taking them in such a back-ass way."[9]

Doc Ewing happened to be in the midst of a professional windfall. At the time, he was running a lab out of a cramped basement on the Columbia University campus. But he would very soon become the founding director of a brand-new earth sciences institution based in a donated mansion on the Hudson River. A former CEO of J. P. Morgan, Thomas W. Lamont, had recently died, and his widow, Florence, had donated the Lamont family estate to Columbia. Doc was intensely interested in sounding the seafloor, and he needed people like Marie, who could draft and run equations and turn seafloor soundings into maps. He gave her a job, and Marie became the first woman allowed to work in science at the brand-new Lamont Geological Observatory in Palisades, New York.

In the pictures from these early days at Lamont, Marie looks serious, with big brown eyes, a small mouth, and carefully coiffed dark hair. One photo shows her seated before a gigantic drafting table with a map of the global seafloor spread out before her. "It was a once-in-a-lifetime—a once-in-the-history-of-the-world—opportunity for anyone, but especially for a woman in the 1940s," she wrote.[10]

She arrived just as Lamont Hall was undergoing a transformation from millionaire's mansion to research institution. The indoor swimming pool became a cafeteria, the greenhouse a machine shop; the kitchen housed a geochemistry lab.[11] Alma Kesner, a longtime Lamont personality and administrator for twenty-five years, remembered that her first office was in a former playroom that had bars across the window. "When the Lindbergh kidnapping happened, the Lamont people had grilles put on all their downstairs windows because they didn't want anybody kidnapping the Lamont children," she said later. "People always asked me, 'Alma, what did you do wrong that you got put in prison?'"[12]

Those were the glorious early days at what is now the Lamont-Doherty Earth Observatory. Back then, Lamont functioned more like a rowdy start-up than the esteemed scientific institution it is today. Friday-evening drinks began at 4:00 p.m. in the grand Lamont living room. The men-only parties in the machine shop went late into the night. Geochemists, geophysicists, and oceanographers ate lunch together, swapping stories, holding informal meetings and talks. In the beginning, there were only a dozen scientists rambling around the estate. Marie drew maps for whoever needed them, but pretty soon one scientist came to occupy all her time.

Doc Ewing had mentored Bruce Heezen after his undergraduate degree at the University of Iowa and taken him on research cruises all over the world; one lasted thirty-six months. Along the way, Bruce learned from Doc and amassed a huge collection of soundings, most of them from the North Atlantic. Bruce also secured funding from Bell Laboratories—the research arm of the communications company—to turn those soundings into maps. Bell was laying a commercial trans-atlantic cable from Newfoundland to Scotland and, more secretly, another cable for the US Navy that would monitor enemy subs.[13] Bell sought Bruce out for his advice on avoiding underwater hazards in the North Atlantic, something Bruce had studied extensively while he was at Columbia University writing his master's thesis on an earthquake and cable break off the Grand Banks of Newfoundland in 1929. The curious thing was that the cables had broken hours *after* the magnitude 7.2 earthquake, and Bruce hypothesized that the earthquake had set off an underwater avalanche, what he called a turbidity current, that had cascaded down the Grand Banks and sliced the telegraph cables as it went. "The speed of the Grand Banks turbidity current is now reckoned to have been as much as 45 miles per hour," wrote the science journalist Robert Kunzig, and its "range was far greater than any avalanche on land."[14]

As she worked under Bruce, Marie's job was at once incredibly simple and quite complex. She took basic depth soundings, known as echograms, and transformed them into readable maps. Along with the soundings collected during Bruce's yearly cruises, she had access to thousands of other depth measurements from ships laying telegraph cables across the Atlantic and from submarine warfare during the world wars. More maps came in all the time.[15]

Surveying the seafloor had gone through a renaissance over the thirty years spanning the world wars, buoyed by the largesse of the US military. In the early 1920s, a US Navy ship collected the first continuous survey line across the Atlantic. With the new setup, a sonar operator, listening via a pair of headphones, recorded the sounding's return and calculated the depth in about a minute. Then he did the calculation again and again at regular intervals across the entire ocean.[16] During World War II, the echo sounder got another upgrade when Doc worked closely with the navy to pioneer the first continuous and automatic sonar. When the sonar sent out a ping, it set a stylus moving across a spooling roll of paper, recording its journey through the depths. After the sounding bounced off the seafloor, a microphone embedded in the ship's hull listened for the return and the stylus burned the depth measurement into the paper with an electric spark.

"The result was an uninterrupted series of seafloor depths along the ship's course. Relatively uninterrupted, that is: The echo sounder depended on the ship's electric power, which went off whenever someone opened the ship's refrigerator," wrote Marie about one problem-plagued cruise across the Atlantic. "When that happened, no echo returned and the sounder recorded depths as bottomless as the crew's appetite."[17] That was the original flyer, the "little lost pings" that Renato Kane worked so hard to clear out on board *Nautilus*.

All of that might sound primitive compared to the multibeam sonars of today, another wartime invention, but it was a quantum leap forward from surveying in the Victorian era, when crews had spent half a day collecting a single depth measurement, lowering a lead line deep into the ocean and then hauling it back up again. Doc was a great, almost rabid collector of data and Lamont became a massive storehouse of seafloor soundings. In the early 1950s, Marie was working with the most cutting-edge knowledge of the seafloor.

One problem with the military's supporting all the seafloor surveying, however, was that data could suddenly become confidential at any moment. Everyone in the main Lamont building had to have military clearance, including Alma Kesner, who ran accounts payable and purchase ordering.[18] Bruce instructed Marie to turn a selection of soundings into six parallel tracks straight across the North Atlantic. She and a team of drafters and human calculators converted the measurements into profiles that look a little like a heartbeat monitor, except rather than tracking heartbeats, they followed the ups and downs of the underwater valleys, trenches, and seamounts.[19] It took the team another six weeks to put the profiles into the right order, west to east across the ocean.

"Plotted on a map, the ship's tracks looked like a spider's web, with the rays radiating out from Bermuda, where most of the research vessels took on supplies and water. Sometimes, the tracks zigzagged, as the ships fled from the paths of storms," remembered Marie. As she stood back to look at her work, a peculiar shape caught her eye. At roughly the same spot in the profiles, a huge ridge rose up from the seafloor, and within that ridge was a little V-shaped cleft. "The individual mountains didn't match up, but the cleft did, especially in the three northernmost profiles," Marie wrote. "I thought it might be a rift valley that cut into the ridge at its crest and continued all along its axis."[20]

As Marie pored over the profiles, she began to believe that the rift might confirm a theory largely dismissed by leading North American geologists. When she showed the map to Bruce, he saw it, too. "It cannot be," he groaned. "It looks too much like continental drift." It was an uncomfortable finding, but Marie couldn't deny what she saw. "If there were such a thing as continental drift, it seemed logical that something like a mid-ocean rift valley might be involved," she wrote. "The valley would form where new material came up from deep inside the Earth, splitting the mid-ocean ridge in two and pushing the sides apart." Now she just needed more evidence to convince everyone else.

3.

Before the Lamont Geological Observatory, before Marie Tharp and Doc Ewing and Bruce Heezen, there was Alfred Wegener, who created the theory of continental drift. Born in 1880, the German meteorologist had clearly spent some time looking at a world map. And like many before him, he noticed that if you removed the oceans, the continents snapped together like a giant jigsaw puzzle. The bulging east coast of South America fit neatly with the tapering west coast of Africa. North America's east coast did the same with the North Sea coasts of Europe. Perhaps the oceans and continents were not eternally fixed, as many people had long assumed, but rather the entire surface of the planet was on the move—albeit very, very slowly.

While convalescing from an injury sustained during World War I, Wegener wrote up his theory in a 1915 book titled *The Origin of Continents and Oceans*—a nod to Charles Darwin's *On the Origin of Species*, which had upended the religious and scientific order a generation before. Clearly Wegener had big ambitions for his new theory. He proposed that the seven separate continents we know today were once a unified supercontinent named Pangaea. He drew from research in a wide variety of fields to prove it—by paleontologists who had discovered fossils of ancient species on both sides of the Atlantic; by geophysicists writing about isostasy and the moving mantle of the Earth's surface—and about glacial deposits that matched up across continents. The evidence was compelling, but so, too, were the gaping holes.

"A page here, a paragraph there, and Wegener, right or wrong, presents enough ill-documented ideas to enrage almost every specialist who was not already enraged by the chapters on geology, paleontology and paleoclimatology," wrote the Scripps Institution oceanographer H. W. Menard.[21] Most American scientists of the day hated the idea of continental drift.

In the late nineteenth and early twentieth centuries, the sciences were trying to professionalize and move away from the gallivanting generalists of the past, including the German naturalist Alexander von Humboldt, who had valued scientific observation.[22] Wegener clearly belonged to that earlier tradition. He broke a world record by staying aloft in a weather balloon with his brother, the polar explorer Kurt Wegener, for fifty-two hours straight. He led expeditions across Greenland, traveling hundreds of miles by dogsled.[23] He wrote poems about his love of tobacco. Intellectually, he was a synthesizer of ideas, someone whose curiosity knew no bounds, whose interests strayed far outside his training, into the fields of astronomy, meteorology, and physics. He wandered so far that he had trouble landing a job in the type of compartmentalized university departments we know today. All this looks rather ironic in retrospect, as Wegener's name now adorns one of the most respected research institutions in Germany: the Alfred Wegener Institute in Bremerhaven.

With the publication of *The Origin of Continents and Oceans*, Wegener veered well outside his lane again. Here was a meteorologist stomping loudly into the fields of paleontology and geology, proposing theories that would upend the life work of some of their most respected scientists. "Wegener's theory was widely discussed in the 1920s and 1930s. It was also hotly rejected, particularly by geologists in the United States who labeled it bad science," writes the science historian Naomi Oreskes. European, Australian, and South African scientists were more receptive to the theory,[24] but in the United States, believing in continental drift was an academic death sentence. Until the continental drift theorists provided a solid explanation for how the continents had done the drifting, it was all circumstantial evidence, no smoking gun.[25]

One hurdle was that much of the convincing evidence of continental drift was covered by the ocean. The first early ocean expeditions had already discovered parts of the rift valley that would turn Marie Tharp into a dedicated "drifter" decades later. In the 1850s, the director of the US Navy's Depot of Charts and Instruments, Matthew Fontaine Maury, discovered a ridge in the middle of the Atlantic Ocean during one of the first systematic ocean surveys. Even with only two hundred soundings across the Atlantic, Maury clearly spotted the rising ridge in the seafloor. He called it a "sea gash"—a term I wish were still in use today—and marveled at the rugged mountain range separating the continents of North America and Europe. "The wonders of the sea are as marvelous as the glories of the heavens," he wrote, "and they proclaim, in songs divine, that they too are the work of holy fingers."[26]

Maury deliberately cast his quest to explore the sea in spiritual language. The study of deep, dark trenches would need a little divine help compared with researching the naturally ethereal sky. Astronomy was one of the most respected fields of the day. Throughout the eighteenth and nineteenth centuries, astronomers had proved the real-world benefits of their work: improving navigation at sea, saving the lives of sailors, and accelerating transatlantic crossing times.[27] Contemplating the cosmos also has a naturally spiritual association, and astronomy has a long history within the Catholic tradition. The Vatican Observatory maintains one of the oldest astronomical institutes in the world, tracing its roots back to 1582. With both economic and religious clout behind it, astronomy outstripped oceanography at the very beginning of publicly funded science in the United States—a gulf that widened into the chasm we see today between the budgets of NASA and NOAA.

Twenty years after Maury discovered the "sea gash," another expedition spotted the rift valley at a different location in the Atlantic Ocean and followed it into the Southern Ocean. In December 1872, a British warship sailed out of Portsmouth, England, on something of a moon shot for ocean science. The HMS *Challenger*'s mission is famous today for having discovered thousands of new marine species on its three-year trip around the world. It hauled up the first seafloor minerals and discovered major geologic features, including the eponymously named Challenger Deep in the Mariana Trench.[28] Over its years at sea, the crew took more than three hundred soundings all over the world, drawn up laboriously from the deep. In the Indian Ocean, *Challenger* discovered a conspicuously shallow stretch of water that would turn out to be connected to the world-girding midocean ridge system. During its return to England, *Challenger* sailed over another section in the Southern Ocean, and the scientists on board speculated that it might extend across the Atlantic all the way to Iceland. But in general, the gentlemen scientists on board *Challenger* were torn over how to interpret the rugged terrain at the bottom of the ocean. Discovery followed discovery, but with such scant information about the ocean available, it was difficult to interpret what they were finding.

The work of hauling each individual sounding up from the seafloor showed an intense dedication to understanding the ocean, to be sure. It also displayed Great Britain's colonial might and the massive resources it could deploy to pursue scientific enlightenment. "The act of inscribing a numerical depth on a chart offered a kind of personal glory and national prestige," writes the historian Helen Rozwadowski about the early ocean surveys of the nineteenth century.[29] The scientists on board vacillated between the major theories of the day. In the long-standing "fixist" view of the earth, land and sea could not simply swap places.[30] Land was new earth; seafloor old. Sediment washed off the newer, more interesting geology on land and came to rest at the bottom of the sea.*

Another popular belief at the time was that the high ridges in the middle of an ocean were actually drowned continents. When *Challenger* discovered the midocean ridge in the Indian Ocean, English newspapers hailed it as the lost city of Atlantis. The myth of a lost underwater city cropped up often in Victorian science fiction: the story of a high-tech civilization wiped out by natural forces held a certain dystopic appeal for readers in the rapidly industrializing England. The public was also primed to accept that other ancient myths might be true. The German treasure hunter cum archaeologist Heinrich Schliemann had recently discovered the lost city of Troy by rambling around Turkish hillsides with a dog-eared copy of Homer's *Iliad* in hand. Why not the lost city of Atlantis, too?[31]

Despite the skeptics, Alfred Wegener remained convinced that the theory of continental drift was true. The rebuke of leading scientists didn't deter him; it seemed to have the opposite effect, goading him to work harder and collect even more evidence. In 1930, at forty-nine years of age, Wegener set out on his fourth expedition to Greenland, where he planned to conduct geophysical experiments on a polar ice cap. During heavy snowfall and the late-autumn temperature of –58 degrees Fahrenheit, he and two companions skied provisions to an inland research station. On the return journey, he and the Greenlander Rasmus Villumsen disappeared.

When a search party found Wegener the following spring, he was still in his sleeping bag, wrapped in furs. (Rasmus Villumsen's body

---

* Another early theory in evolutionary biology followed similar thinking: that deep-sea animals had evolved from animals living in the shallows that had migrated downward. We now know that the opposite can be true, that life can migrate up as well as down.

was never recovered.) They concluded that Wegener had likely died of a heart attack in camp. They also reported that the dead man looked strangely happy, a beatific smile frozen on his face. He died never witnessing his maligned theory become his greatest contribution to science, but perhaps he had guessed that one day he would be proved right.[32]

4.

Marie Tharp didn't trek across Arctic tundra or ride in hot-air balloons like Alfred Wegener, but she was an outsider in her own way. As a divorced woman working in the insular and mostly male scientific world of the 1950s, she struggled against restrictions big and small, spoken and unspoken. Perhaps the most egregious was that she wasn't allowed to collect her own soundings at sea. All the American marine science institutions—Scripps in La Jolla, California; WHOI in Woods Hole, Massachusetts; and Lamont in Palisades, New York—forbade women working on research vessels until the late 1960s. At Lamont, they weren't even allowed to set foot on a gangplank.[33] The prohibition stemmed from an ancient superstition among sailors that women brought bad luck and disaster at sea,[34] although the official reasons were that the ships had not been built to accommodate both sexes and that men would have to supervise women on board. Research vessels during the 1950s and '60s were in fact much more dangerous places than they are today. Back then, it was standard practice to sound the seafloor by throwing a stick of dynamite overboard every two minutes. A researcher on board an early Lamont ship was blown apart and buried at sea.[35] During a gale off Bermuda, a rogue wave swept Doc Ewing and three other men overboard. The captain managed to recover three of the four men, including Doc, who walked with a limp the rest of his life.[36]

The obvious solution would have been to improve safety on board, and Lamont did phase out the use of dynamite and instead developed seismic air guns that could be towed behind the ship. But the vessels still operated with an abundance of machismo, adopted from the navy and nautical worlds. "Part of the issue is an emotional one. For many men going to sea represents a temporary reversion to boyhood, and they don't want their glorious chance to get away from it all endangered," said one mathematician at WHOI at the time.[37] As Marie was unable to collect her own data, her insights could always be dismissed as the work of an armchair geologist without real-world experience.

One way around the restrictions was for a woman to find a male scientist to act as a patron. That was the root of the relationship between Bruce and Marie. Bruce went to sea and collected data; Marie analyzed and interpreted the soundings back on land. That wasn't such an unusual arrangement in oceanographic circles at the time. "These men considered it glamorous and pleasurable to go to sea, far more so than staying at home to analyze [data]," writes the science historian Naomi Oreskes. "This is one reason data analysis was often left to women."[38]

Tracing their trajectories through the academy is a good reflection of how men and women fared at the time. When Marie arrived at Lamont, she had three degrees under her belt, along with work experience drafting for an oil company and the US Geological Survey. Even though Bruce was four years younger than Marie and had only a master's degree when they met, she became a research assistant to Bruce, who was still a PhD student. It took him nearly a decade to finish his doctorate because there was simply no incentive. "He already had a job, money, a supporting staff for research and was leading expeditions. What did it matter whether he had a Ph.D.?" wrote the Scripps oceanographer H. W. Menard, a close friend of Bruce.[39] Even before becoming a senior in college, Bruce had commanded his own research vessel. He had become chief scientist on his second research cruise, which he took around the time he finished his undergraduate degree.[40] He won awards and tenure at Columbia University, making him impossible to fire—something that would prove crucial later in his and Marie's lives.[41] Meanwhile, Marie worked one tenuous Lamont contract to the next, her name often scrubbed from publications.

All that made no difference to Marie as long as she got to do what she called "important work." "I thought I was lucky having Bruce as a boss because he gave me such a challenging job," she said. "And I didn't care what my classification was or anything. Assistant, draftsman, computer assistant. I didn't care what it was. Because I felt that I was working on the same problems he was. I didn't have any resentment."[42]

But some women did resent the sexist rules, particularly the one against going to sea. That rule stymied the careers of an unknown number of aspiring female oceanographers, who were stuck on land and unable to collect their own data. In 1955, a graduate student circulated a pamphlet at Woods Hole condemning the policy. "Some scientists have said that it isn't necessary to go to sea in order to be an oceanographer," wrote Roberta Eike, a young biology graduate student at WHOI, "but I would prefer to collect my own data and to have the opportunity to make all the important personal observations that go with it."[43]

A year later, with no change in the rules, Eike took matters into her own hands and stowed away on a marine biology cruise. When her supervisor, George Clarke, discovered her on board, he held her over his knee and spanked her. The ship made a U-turn and dumped her back on shore, where Woods Hole canceled her fellowship.[44] Nothing is known about what happened to Eike after her expulsion from WHOI.

By that time, Marie was in her late thirties, and she might have felt a little past that kind of youthful rebellion. She had started at Lamont nearly a decade earlier, when it had still been rare for any woman to have a position on the scientific side of things. Still, she felt unwelcome, even by the man who had hired her. "I think he hated their guts," she said of Doc Ewing's feelings about women in science.[45] She turned her attention to the drafting table instead. On the West Coast, two women at the Scripps Institution of Oceanography did the same, drafting and plotting seafloor maps for their researcher husbands.[46]

Geography is one of those odd gray areas of study. The subject straddles the divide between the traditionally feminine pursuits of the arts and humanities and the traditionally male domains of exploration and math. In the early days of the United States, geography was one of the few scientific subjects taught to women. Geography is "a study so universally instructive and pleasing, that it has, for nearly a century, been taught even to females," wrote John Pinkerton in a preface to an 1818 atlas.[47] There were variations in how the two sexes learned geography, however, with courses for women focused heavily on painting, drawing, and even stitching together maps of the United States. The aim was partly patriotic, burning the boundaries of the young republic into the minds of all Americans regardless of their sex. That early cartographic tradition faded as maps became easier to print, but it opened the door, however slightly, to women in science and might have allowed Marie to practice science under the guise of a more feminine craft.[48]

Still, Marie struggled to fit in at Lamont. Marie rarely mentioned her marriage to a violinist in Ohio, which failed before she joined Lamont,[49] lest she become even more ostracized by the male scientists who had wives and children at home. A student of Bruce, the oceanographer Bill Ryan, remembered that Marie had the "mannerisms of a little girl, giggling and talking in a squeaky voice."[50] Well into her seventies, she referred to herself as a girl. Her casual style also raised eyebrows. Before black-tie events at Lamont's parent institution, Columbia University, Bruce would dispatch Alma Kesner, the administrator, to review Marie's outfit so she wouldn't show up in a long gown and sneakers.[51] "Marie had a certain way of dressing that nobody could really imitate," said Kesner. "But today when you look around at people walking the street today, they all look to me like Marie."

For all her contortions to fit the strictures of the day, Marie didn't compromise on the science. When Bruce saw her early maps of the Atlantic midocean ridge and heard her expressing dangerous ideas about continental drift, he rejected her interpretation as "girl talk."[52] "At the time, believing in the theory of continental drift was almost a form of scientific heresy," she wrote. "People who believed in continental drift even had a derogatory name within the community; they were known as 'drifters.'"[53] Promoting continental drift could get Marie and Bruce cast out of the scientific firmament, particularly in the United States during the 1950s and especially at Lamont, where Doc Ewing was a devoted fixist.[54]

It took Marie about a year to convince Bruce that the ridge cutting through the Atlantic was real. Two pieces of evidence eventually won him over. A deaf art school graduate sitting next to Marie had arduously hand plotted thousands of underwater earthquakes for the Bell Laboratories map. The earthquakes lined up right along Marie's rift valley, proving that the ridge in the Atlantic seafloor was geologically active. Another convincing point was that geologists confirmed that the East African Rift, slicing through Djibouti, Ethiopia, Kenya, and Tanzania, was a slowly spreading crack in the Earth's crust. (Most of the midocean ridges are underwater, but a few, such as the East African Rift, run up onto land.) Finally, Bruce accepted Marie's interpretation, but he couldn't bring himself to support continental drift. "It was very hard to go in the direction of that theory when the boss, Doc, like nearly everyone else in the scientific world, was violently opposed to drift," she wrote.[55] Instead, Bruce pushed an alternative theory: the Earth was slowly expanding, driven by molten magma pushing up between the rifts in the seafloor.

It took another four years to publish their findings on the midocean ridge system, as the rift came to be known. In the intervening years, Marie dived into the seafloor soundings collected during research cruises and used them to extend the midocean ridge system all over the world. She pulled soundings from *Meteor*, the 1925 German research vessel that had collected the first set of evenly spaced transatlantic soundings across the Southern Ocean and discovered the South Sandwich Trench. Buried in the unanalyzed soundings, she spotted the midocean ridge in the V-shaped indentation yet again, extending the mountain range from the Atlantic Ocean into the Southern Ocean just as the scientists aboard *Challenger* had predicted nearly a century before. She culled soundings from a Danish expedition that had taken place two years after *Meteor* and found a similar ridge in the Indian Ocean, named Carlsberg Ridge after the famous Danish brewery and scientific foundation.[56]

When Bruce began to present the work in the late 1950s, he took a globe with him to show scientists the ridge's path around the planet. The reaction ranged from amazement to disbelief to outright hostility, according to Marie. After one of Bruce's presentations in 1957, the Princeton University geologist Harry Hess stood up and declared, "Young man, you have shaken the foundations of geology." The famous ocean explorer Jacques Cousteau was among the skeptics, but he had a way to test the theory. On a trip across the Atlantic, he strapped a video camera to a sled and towed it behind his ship, right over the midocean ridge. In 1959, he screened the footage during an international ocean-ographic conference at New York's Waldorf-Astoria hotel. "They were beautiful movies of big black mountains with white snow drifts and blue water," Marie remembered, watching transfixed from the audience. "And he took photographs of it, and so that helped a lot of people believe in our rift valley at a time when a lot of people doubted it."[57] The footage of the underwater mountain range won over more skeptics.

Harry Hess, the Princeton geologist who had lauded Bruce's presentation in 1957, had set to work explaining continental drift. A naval commander during World War II, Hess had made sure to leave the sonar running at all hours to gather as many soundings as he could across the Pacific. He would go on to develop the seafloor spreading theory, an explanation of how the continents drifted along the Earth's mantle. Seafloor spreading follows the same rolling motion you see in a boiling pan of water, as the science journalist Robert Kunzig explained. Hot, molten earth rises from cracks in the seafloor and pushes aside the older, colder seafloor in front of it. This convective motion works like a conveyor belt, shuffling old seafloor farther away from the midocean ridge until it either runs into the continental shelf or becomes so old and dense that it sinks into an ocean trench, from which it is sucked back into the Earth's mantle.

The seafloor spreading theory kicked off one of the most exciting breakthrough periods in the study of the planet. Nearly every year after Hess published his idea, a new theory emerged that supported continental drift, often by scientists making separate, simultaneous discoveries. At the time, the geological community was puzzled by odd magnetization patterns radiating outward around ocean ridges. Then three geologists—Drummond Matthews and Fred Vine at Cambridge University and Lawrence Morley, a geophysicist with the Geological Survey of Canada—separately proposed a hypothesis for validating seafloor spreading using the zebra-striped magnetization patterns. If Hess's theory held true, new magma pools up from the ocean ridge, splits in two, and cools. The highly magnetic basalt of the new seafloor aligns with the Earth's current polarity, effectively time-stamping the seafloor, because the Earth's magnetic field reverses every 300,000 years or so, with the magnetic north becoming the magnetic south and vice versa. After each reversal, the next batch of magma records the 180-degree switch in magnetization. If you know how to read the alternating strips of magnetization, they're a clear catalog of seafloor spreading in action.[58]

Two more papers by the Canadian geophysicist John Tuzo-Wilson[59] followed in quick succession, addressing the remaining holes in the concept of continental drift. Tuzo-Wilson outlined a theory of why tectonic plates do not all behave in the same way: some plates slide under, over, across, or away from each other. On land, we experience just a fraction of that action at the rare meeting points between tectonic plates. The San Andreas Fault is one of the best examples, running onshore near Eureka, Oregon, and slicing under much of California before tapering off in the desert mountains outside Palm Springs. The San Andreas Fault divides the North American and Pacific plates, but it's a transform fault, meaning that the two plates rub along each other in a north-south direction, releasing little seismic shudders that I can feel in San Diego. The windows rattle, the bed frame shakes, and I pray that the big one will hold off just a wee bit longer.

At the Mid-Atlantic Ridge, the plate boundary behaves differently once again. The plates here are moving away from each other at a rate of around two inches each year.[60] New molten rocks burble up from the Earth's core to fill the gap; that rock cools into seafloor and then is shuffled along the conveyor belt of the ocean floor. In other words, the Atlantic seafloor is expanding—very, very slowly. In the Pacific, the oldest ocean in the world, the opposite is happening. The Pacific is contracting as old seafloor slides down into the trench system along the Ring of Fire and is sucked down into the Earth's mantle. The midocean ridge system is like the bursting seams of the planet. Or, as Bruce Heezen so poetically put it, it is "the wound that never heals."

Tuzo-Wilson tackled another troubling problem with plate tectonics: How can volcanoes form so far from the midocean ridge? Hot spots in the Earth's mantle create volcanoes as the tectonic plates move over them, producing volcanic islands such as the Hawaiian Islands chain. After Tuzo-Wilson, two researchers—Dan McKenzie at the Scripps Institution and Jason Morgan at Princeton University—published separate papers in 1967 and 1968, proposing that the Earth's surface was covered by a shifting patchwork of rigid plates, what we now call the plate tectonic theory. At last, the final puzzle piece fell into place. For the first time, a single theory could account for all the major features on Earth, from the shape of continents and oceans to the existence of earthquakes and volcanoes. Dozens of scientists contributed along the way, far more than just the big names mentioned here, forging "the first global theory ever to be generally accepted in the history of earth science," writes the science historian Naomi Oreskes.[61]

Although there was tremendous excitement in the scientific community over the discoveries, the general public was still largely in the dark. It was Marie Tharp's map that revealed the great secret of the seafloor to the masses.

5.

During her first years at Lamont, Marie Tharp drew traditional topographic maps of the seafloor. With their wavering contour lines and precisely labeled elevations, those maps are mostly academic documents, dense with information and tricky to read. Soon after she arrived at Lamont, she wasn't allowed to publish them. In 1952, the Pentagon classified a broad swath of earth science information, from the upper limits of the atmosphere all the way down to the shape of the seafloor. The major Cold War projects at the time involved ballistic missiles and antisubmarine warfare, which depend on gravity measurements and the shape of the seafloor. Any seafloor measurement greater than 1,800 feet became confidential—a matter of national security. She and Bruce either would have to wait years, maybe decades, to publish their work or could find another way to tell the world about the midocean ridge system they had discovered at the bottom of the Atlantic.

Working late into the night at Lamont, as they often did, Marie and Bruce hit on an idea for circumventing the Pentagon's ban: Why not draw the seafloor in a more realistic style? The result would be more general, less scientific, but it would communicate the spirit of the seafloor and prove it was not a flat dumping ground but a diverse, rugged terrain. Bruce took a pen and started to sketch the western

Atlantic seafloor from memory. He drew the shoreline, the descending continental slope, the abyssal plains, the plateaus, and the ridges. It took him about an hour to finish the preliminary sketch, and then he handed it to Marie. "Well, why don't you fill in the rest?" he said. Neither of them realized that he had just given her a task that would consume the rest of her life.

Marie began to draw in a looser physiographic style that showed the seafloor at an oblique angle, the way the Rocky Mountains look through an airplane window on a transcontinental flight. It took all her geographical and geological training to translate the sparse data points into more understandable terrain. On land, geologists climb a mountain, look around, take measurements, and make a map. Marie didn't have the opportunity to survey the seafloor with her own eyes; she had to decide what features to emphasize and create the "feel" of the new frontier rather than a set of recorded data points.[62]

"It was a very demanding technique where you had data. You could show everything. And it allowed you to invent and extrapolate where you had no data," she said.[63] She needed that interpretative wiggle room because there was so much blank space to fill in. In the Atlantic, one of the best-mapped oceans in the world, the survey lines fell on average 130 miles apart. In some sparsely mapped regions, she used measurements of underwater earthquakes to predict as best she could where the ridge might be. "They had vastly more sounding data to work with than Matthew Fontaine Maury or any other cartographer, but given the size of the ocean, the amount of data they had was still ridiculously small," wrote the science journalist Robert Kunzig. "To say that they relied on intuition in sketching the seafloor is to engage in euphemism: they made most of it up."[64]

Cartographers have always had to take some creative license with charting the unknown, or else no maps would ever have been made. Up until the eighteenth century, mapmakers simply scrawled "Terra Incognita" or "Terra Nullius" across vast stretches of map and were done with it.[65] That was at least honest compared to what even earlier cartographers had done in the Middle Ages, unleashing their imagination upon unknown continents. "Early maps of Africa and Asia are veritable zoological gardens superimposed on the earth, with animals and fishes gamboling over vast savannahs of geographical ignorance," writes Stephen Hall.[66] In one poorly reconnoitered part of the North Atlantic, Marie did the same. Borrowing a technique from the portolan charts of the thirteenth century, she suggested adding a few mermaids and sea serpents to the map. Bruce would have none of it. They compromised by placing a large legend over the top of one blank spot—another trick of cartographers past.[67]

They called their new creation a "physiographic map"—a term taken from terrestrial cartography. The first one had appeared in a Bell Labs technical journal in 1956, but it had gone mostly unnoticed by the scientific community until a couple years later, when a geological journal had republished it. That publication had reignited the long-simmering debate about continental drift and set into motion the breakneck pace of discoveries in geology throughout the early 1960s. Bruce showed an early draft of Marie's map to the editors at *National Geographic*, who liked it and hired Marie as a consultant. Over the coming decade, she traveled back and forth to Austria to work with the alpine painter Heinrich Berann, who turned her physiographic maps into vivid painted maps for *National Geographic*. One of the biggest challenges was simply rustling up enough soundings to keep Berann busy. "We'd plot all the data we had, and then there'd be blank spots," she said. "So we'd come back here and work up that spot with what we could get. Then we'd go back again, and Berann would paint it."[68]

They finished the first map of the Indian Ocean right on time. *National Geographic* published it as a supplement and tucked it between the pages of the October 1967 issue—the same year that plate tectonic theory spilled into the mainstream. The map's publication paved the way for continental drift's long overdue acceptance in scientific circles, but the broader public loved the maps, too. The gigantic midocean ridge system neatly divided the Earth into separate plates and made the subtle, mostly invisible movements of plate tectonics plain to the layperson. "Everyone could see it. Everyone including the thirteen million people reading the *National Geographic*. . . . Somebody looked at them with their eyes open," remembered Marie.[69] At another point she wrote, "There's truth to the old cliché that a picture is worth a thousand words and that seeing is believing."[70]

6.

A slim package arrived on my doorstep in San Diego. I took a knife and carefully cut the tape sealing the manila envelope. Then I paused for a moment, trying to picture what a *National Geographic* subscriber in 1967 might have felt, someone who had never seen a complete map of an ocean floor before. I slid the mint condition map out of its package and unfolded it, 19 inches long by 25 across, and laid it on my kitchen table.

At a quick glance, the map shows the Indian Ocean, but with one major difference: the ocean is gone, drained away, disappeared. The seafloor now appears to be a space where someone might go for a stroll. Judging by the pleasant light, it looked to be about midafternoon at the bottom of the ocean. Thanks to some clever shading, the water is invisible but still felt. The terrain on land is colored taupe yellow, coastal water cerulean blue, and the deep sea dark teal. You can see the split-screen effect best at the islands, where taupe-yellow land pokes above the waterline and becomes cerulean down below. The map gave geologists a view of the planet they had hungered to see for decades. With the water drained away and the surface and texture of the entire planet made plain, the view is transcendent, allowing mere mortals to see a place we were never allowed to see. Suddenly, you understand that an island is not land surrounded by water but the peak of a vast underwater mountain range hidden beneath the surface. Each region of the sea has its own character, too. The deep abyssal plains of the Indian Ocean look cold and forbidding, like frigid Siberian hinterlands. The shallow South China Sea looks warm and inviting, a place where I might take a vacation at the bottom of the ocean. Such a map might look at home hanging over the bed of an aspiring ocean mapper or in a university hallway, as many certainly were. It was an immediate smash-hit sensation. More than sixty years later, the *National Geographic* supplement is still the most widely recognized map of the seafloor today and remains in wide circulation.

The discovery of the midocean ridge brought fame to the Lamont Geological Observatory. Alma Kesner, the Lamont administrator, remembered fielding calls from the public after the map's publication.

"People used to call up and say; 'Now this Heezen, you work with him, don't you?' I said, 'Yes.' 'They say there's a rift right around the whole earth. Will you call and ask him if it can split in half.' 'Oh,' I said, 'Yes, I'll ask him, sure,'" she said. ". . . You know, we'd laugh about it."[71]

The Lamont Geological Observatory was the underdog sibling of Woods Hole on the East Coast and Scripps on the West. In less than twenty years, Doc Ewing had transformed the Lamont estate into a world-class observatory that could compete with the big dogs. But with that success had come trade-offs. The convivial atmosphere of the early days faded, and a more civil professionalism took. "Times change; by the end of the [1960s] Lamont had many more scientists, much more money, and an enlarged administrative structure," wrote the Scripps oceanographer H. W. Menard.[72] There were fewer shared luncheons, talks, and parties between research groups. It was natural, even necessary, for the heady start-up culture to cool, but when Doc divorced his second wife and married his secretary, the tension at Lamont ratcheted up. His new wife, Harriet, continued to work as his secretary and ended her husband's long-standing open-door policy. Before, researchers could stroll in for a chat with Doc, but now everyone had to make appointments through Harriet. She quickly became the bête noire of Lamont, and many credited her arrival with the end of the more freewheeling early period at Lamont.[73]

As Bruce's success grew, so, too, did a rift between him and Doc. The definitive rupture happened in 1964, when Bruce declared that Doc's name would no longer appear on his papers.[74] It's standard practice for the name of a laboratory's lead investigator to appear on every paper published by that lab, but Bruce had long chafed against the policy, particularly what he perceived as Doc's taking more credit than was due for his work. In retaliation, Doc limited Bruce's access to Lamont ships, then cut off his access to the shared Lamont data altogether. Still Bruce found a way to get his hands on the records. "Marie Tharp smiled when she mentioned 'mid-night requisitions' during this period," wrote Menard. "Still, Bruce could not publish the data."[75] Doc couldn't tame Bruce, nor could he fire the tenured professor, but he could fire Bruce's longtime research assistant, Marie.

The discovery of the midocean ridge and the publication of the

Tharp-Heezen maps in *National Geographic* should have been Lamont's crowning achievement. Instead, the staff split in two, with one camp supporting Doc, the other Bruce. "I never questioned [Marie's firing] because I was, believe it or not, I was on the side of Bruce," said Alma Kesner. "A lot of [people] said Bruce brought it on himself with his tantrums, you know, like a little boy. And he said, as a result of it, Marie was the goat. She was thrown off the campus completely."[76]

Unsurprisingly, researchers who had been with Lamont since its inception started to head for the doors. Bruce continued to pay Marie's salary through separate navy contracts, and for the rest of her career, she worked out of her house in Nyack, New York, converting nearly every room into a workspace. Up to a dozen people at a time might be huddled over a sprawling map of the seafloor at Marie's house. The wandering only child of her youth, the perennial outsider, had been cast out again. But Marie credited her new work-from-home situation with her being able to complete her incredible task. "As long as Bruce arranged to pay my salary, I had a place to work and people to work for me, I just kept going," she said. Always a workaholic, she now had nothing else to distract her from drawing maps.[77]

Throughout the 1960s and into the 1970s, *National Geographic* published more and more Tharp-Heezen maps, drafted by Marie and painted by Berann. First came the map of the Indian Ocean, the one I held in my hands. Then the Arctic, the Atlantic, and finally the whole planet. In the 1960s, the navy declassified the seafloor soundings, but Bruce and Marie stuck with their original drafting style anyway.

In his later years, Bruce turned his attention toward developing nuclear submarines with the navy, and in 1977, he got the chance to board one and visit a section of the midocean ridge himself. The cruise off Iceland included a stop first in Paris to pick up the world-famous Jacques Cousteau. Before leaving Lamont, Bruce strolled into Alma Kesner's office to say good-bye. "I am going to meet Cousteau. We're going aboard his submarine," she remembered Bruce telling her. Kesner immediately had a bad feeling about it.[78]

"Bruce, come over here to me," she said to him. Bruce came over. He was a stocky, round-faced man who drove himself relentlessly and rarely took vacations. Not someone whom you would call the picture of health.

"What's the matter?" Bruce asked Kesner.

"Do you know how fat you are?" she asked him. "You can't, you won't get through a door, porthole of that little, tiny little thing." More to the point, Bruce had already suffered a heart attack nearly twenty years earlier, in 1959, and ended up in hospital for three weeks. As soon as he had gotten out of the hospital, he had gone on to present thirteen papers at the International Oceanographic Congress in New York and change a flat tire on his way home.[79] That same year, his father had died of a heart attack. Bruce brushed Kesner off, told her not to worry. On his trip to Paris, he took with him the proofs of *National Geographic*'s world map.[80] But he would never get to see that final map published. He died of a heart attack on board the nuclear submarine.

Marie was devastated by Bruce's death. Bruce had never married or had children, and, apart from her first failed marriage, neither had Marie. It was an open secret that the pair were something more than a professional partnership but something less than a romantic one. "She really loved him," remembered Kesner. "And he loved her, too, in a way. In a way that, it was the funny—you know, get off my lap, but I love you anyway. Or get up off that chair. You know." Kesner used to tease Marie about the relationship, asking her when the two were going to get married. "Oh, he's not interested at all," Marie would reply. They talked about it, joked about it sometimes, but nothing ever happened.[81]

After Bruce died, Marie suffered professionally, too. For the first time in thirty years, she was working without his protective patronage, and she began to lose projects she loved. One was drawing maps for the General Bathymetric Chart of the Oceans (GEBCO), the organization behind today's Seabed 2030. Bruce was a long-standing member and editor of GEBCO, which was how the project had come into Marie's hands. "They produced the best maps I used to sketch on," she said of GEBCO, which she called a bootleg project. Although the organization never traded in illicit maps, it did function a little like a bootleg operation, with a small group of die-hard volunteers collecting and sharing as many soundings about the seafloor as possible despite the classified nature of their work. Such a precarious, underfunded task was particularly vulnerable after Bruce passed. "I didn't have a program for [GEBCO] with a contract and money that is. I couldn't finish this work that Bruce started because, after he up and died, they took all his assignments and gave them to other people. And not to me. They came to my house and took the materials away," she said. "That was a very unhappy period in my life, but it was just the beginning of what life was like without Bruce."[82]

The fame of the Tharp-Heezen maps may even have backfired and ended up hurting hydrography unintentionally. The casual viewer looks at the maps today and assumes that the seafloor is charted, the job is done, and we can move on to other things. This is the challenge that Cassie Bongiovanni faces today, more than a half century later, having to explain again and again that the ocean is *not* mapped, no matter what the maps might show. "This is a *characterization* of what [the seafloor] would look like if you could remove the water. It gives you the false impression that it's a map," said Robert Ballard, the founder of the Ocean Exploration Trust and the owner of E/V *Nautilus*, in his 2008 TED Talk, displaying the Tharp-Heezen world map. "It is not a map."[83]

Maps occupy an authoritative position in society and hold a false allure. They trick us into thinking we know a place better than we do, particularly in remote territory. You can't use the Tharp-Heezen maps to find a specific place on the seafloor, as you would an ordinary map. In 1984, a group of oceanographers tried to do that, looking for parts of the rift valley in the Southern Ocean. They found it 150 miles away from where the Tharp-Heezen maps said it should be.

Instead, the Tharp-Heezen maps are more like the maps made by European cartographers at the end of the fifteenth century. One in particular, by the Florentine mapmaker Henricus Martellus Germanus, is credited with leading Christopher Columbus to sail west across the Atlantic, looking for a route to China and "discovering" the Americas instead. Considered cutting edge when Columbus set sail in 1492, the Martellus map shows Earth without the landmasses of America, Antarctica, and Australia. The coastlines of Europe, Africa, and Asia are there, incorporating new knowledge gained from the journeys of Marco Polo and Bartolomeu Dias, but the Pacific, the largest ocean on

Earth, is not much more than a bit of blue at the margins. The huge gaps in the Martellus map suggest that a faster route to Asia might be possible by sailing west. Today we scoff at Columbus, who insisted until his death that he had been in the East Indies when in reality he had gone ashore in the Caribbean. But that was the world he knew based on the best maps of the day.[84] The Tharp-Heezen maps do the same; they freeze-frame exploration at a moment in time, holes and all. I hung the *National Geographic* map of the Indian Ocean by my desk—a reminder that a map is always the start of something, not the end.

Marie spent the rest of her life running a map distribution business out of her house in Nyack, New York, and publishing articles about Bruce. In interviews conducted thirty years after his death, she still spoke of Bruce as though he were sitting there beside her: "Now Bruce could give you some more profound observations than I could," she said in one characteristic aside, still deferring to her longtime boss.[85] Marie often struggled to support her work drafting new maps of the seafloor. No matter how popular the Tharp-Heezen maps of the seafloor became, they were eclipsed by the Space Race, the mania for all things moon shot that climaxed in 1969 when Apollo 11 landed on the moon. The world had moved on to a new frontier, and public investment in ocean mapping had dwindled.

Near the end of her life, Marie began to gain more recognition for her work, and she welcomed curious historians, journalists, and writers into her living room. Despite the tragedies and setbacks, she sounded upbeat in those last interviews, thankful for what she had achieved. "Establishing the rift valley and the mid-ocean ridge that went all the way around the world for 40,000 miles—that was something important," she said. "You could only do that once. You can't find anything bigger than that, at least on this planet."[86]

CHAPTER 5

THE LONELIEST OCEAN ON EARTH

1.

The waves rose and the temperature fell as *Pressure Drop* sailed into the Southern Ocean. A few days earlier, on January 24, 2019, the ship had left Montevideo, Uruguay. After a brief stopover at the island of South Georgia to pay respects to the grave of the great British explorer Ernest Shackleton, the Five Deeps team continued their journey to the world's only subzero hadal trench. It would take roughly five weeks to cross the ocean and reach Cape Town, South Africa, on the other side. A German ice pilot accompanied the crew on board, guiding the ship through the iceberg-strewn seas.[1]

The Atlantic, Pacific, Indian, and Arctic are the obvious oceans, if you will, clearly defined basins hemmed in by continents. Many people would struggle to name the fifth and final ocean. I've heard it called the Antarctic or South Atlantic Ocean, the South Pacific or South Indian Ocean. There is in fact no internationally agreed-upon name, although scientists have called it the Southern Ocean for more than a century.[2] Whatever you call it, just know that the Southern Ocean is the exceptional ocean, swirling unfettered around the icy white continent at the bottom of the planet. The Southern Ocean's border is defined by the swift clockwise current encircling Antarctica. There is no landmass to interrupt the current, so the waves build to exceptional speed, strength, and height.[3] The highest wave ever recorded there was as tall as a seven-story building, but because the environment is so brutal and difficult to monitor, that's just an estimate. There's something magical about a place where waves as high as skyscrapers circle the globe unseen by human eyes.

The harsh conditions and the length of the journey had caused some anxiety among the crew. One Triton Submarines staff member suffered a panic attack before leaving Montevideo and had to take a day of bed rest before being cleared to sail. Geographically, the Five Deeps team was heading into a stretch of ocean as far removed from civilization as one could possibly get without blasting off into space. Just north of the 60th parallel south is Point Nemo, the point that is farthest away from land in all directions. Named after Captain Nemo, whose Latin name translates to "no one," Point Nemo is more than 1,450 nautical miles away from Ducie Island, part of the Pacific Pitcairn Island chain and the refuge of the *Bounty* mutineers, Motu Nui in Chile, and Antarctica's Maher Island. There is no marker or buoy at sea to announce Point Nemo. It is literally nowhere: just a couple coordinates on a map. The Southern Ocean feels just as remote as Point Nemo—cast off from the world, nothing but miles and miles of water in all directions. Until the ship reached Cape Town on the other side, the forty-four people on board would be all one another had.

So where is the Southern Ocean exactly? Because there is no landmass to mark precisely where the Atlantic, Pacific, and Indian Oceans end and the Southern Ocean begins, there is debate among scientists, politicians, and sailors alike on where to draw the border. "That boundary will change if you ask oceanographers, chemists, biologists," explained Aileen Bohan. "As the seasons change, biology changes, the chemistry is changing with the weather, and the oceanography is changing with the currents and with El Niño. Then there are different political boundaries, too."

Bohan and Cassie ended up going with the 60th parallel south, the boundary set by GEBCO and the 1959 Antarctic Treaty, which governs the continent. That was an inconvenient choice for Five Deeps. The South Sandwich Trench straddles the south 60th latitude line dividing the two oceans. Worse, the Meteor Deep, widely believed to be the trench's deepest point, falls to the north of the 60th parallel south, meaning that Victor would be diving a shallower point in the same trench based purely on one definition of where the Southern Ocean is. What if some authority back on land defined the Southern Ocean differently and refused to recognize Victor's world record? Just to be safe, the expedition aimed to dive the deepest points on both sides of the 60th parallel south in the South Sandwich Trench.

As they sailed, dolphins leaped against a backdrop of blinding blue-white icebergs. "We have been dodging icebergs all day and just passed the southernmost islands in the South Sandwich Islands chain: Thule and Cook Islands. Volcanic islands both, we saw wisps of steam curling up from their summits and cold lava flows on their flanks with icebergs just offshore," recounted Victor on his expedition blog.

The Southern Ocean started to batter the Five Deeps team, tucked inside the dry lab. Pens and paper went flying. Someone had made the unfortunate choice of stocking the lab with rolling chairs, which careened around the heaving room like bumper cars. A few people stuck duct tape into the gimbals at the bottom of the chair legs, hoping to slow their roll—with dismal results. After a particularly steep wave, people called out bets to Cassie and Bohan, seated at the sonar desk watching the live data stream in. "Twelver!" someone shouted as *Pressure Drop* heeled a hard twelve degrees to one side.

As soon as the ship arrived over the southern tip of the South Sandwich Trench, the clock started ticking for Cassie and Bohan. Bohan took the night shift, running survey lines back and forth over the trench. In the morning, Cassie took over, interpreting the data, writing reports, attending meetings. The two mappers stayed glued to the sonar desk, trying to catch and clean errors as soon as they appeared. Every measurement, whether it's taken above the waterline or below, comes with some amount of uncertainty. That was the hardest thing for me to grasp about an ocean mapper's job. As you stand on terra firma, ruler in hand, the principle can be hard to wrap your mind around. One inch equals one inch equals one inch, right? But as I saw on *Nautilus*, errors have a way of creeping into any measurement. The more extreme the landscape, the more the uncertainty increases and the statistical equations for predicting it change. Pitching and heaving on the Southern Ocean are at the absolute extreme ends of surveying the earth. Errors definitely creep in.

"Victor is very nice and friendly, but he's also like, 'I need to know where the deepest point is, and I can't hear in a year that it's wrong,'" said Bohan.

Before *Pressure Drop*'s visit, the majority of the Southern Ocean was uncharted; the same can be said for the South Sandwich Trench. With 91 percent of the trench unmapped, the existing maps had been drawn mostly from satellite predictions. The EM 124 transformed the satellite-predicted blur into a sharp three-dimensional terrain of ripples, cracks, and tears in the seafloor. The contrast between the known and unknown was so striking, it felt like putting on a pair of prescription glasses for the first time. Victor was so excited that he printed out the new maps and posted them around the ship. "I think we were able to show to Victor what mapping was," said Cassie. "Most of the people on board had never seen [bathymetric] maps or really knew what mapping was about, and we were able to prove its value for that trip."

Not long after the ship arrived over the South Sandwich Trench, the Scottish captain and the German ice pilot spotted a calm weather window approaching. The best day for the dive was determined to be the morning of February 4, 2019. Even after a full day at the sonar, Cassie was too excited to sleep. She decided to stay up with Bohan as she ran the last survey lines over the trench. She grabbed a stack of dinosaur sticky notes off her desk and passed them around the dry lab. "Make your bets, everyone," she announced. There were three candidates in the region that might hold the deepest point. The ship's crew were about to make history that night: they were going to find out which one had the deepest point in an entire ocean.

Later that evening, "Cassie and I were just sitting there behind the computer," remembered Bohan, watching the numbers roll in, and then suddenly, the basin that no one had expected to hold the deepest point revealed a superlative number.[4] It was a quiet, almost anticlimactic moment, as so many moments of discovery are. There was no shouting of "Eureka!" or waking up the rest of ship. It was just the two mappers, alone in a lab. They looked at each other. "And we're just like, 'That's it. Jesus. What do we do now?'" said Bohan.

An electrical engineer had picked the winning number. The Southern Ocean's deepest point was 7,434 meters (24,389 feet), give or take

13 meters (42 feet).[5] The next morning, just a few hours after they discovered the deepest point in the Southern Ocean, Victor Vescovo would dive it.

The Southern Ocean is special in another way: it's one of the last marine wildernesses on Earth, where more than 50 percent of the ocean is still untouched by human impact. As Victor sank down into the Southern Ocean, he was dazzled by the marine life passing by the viewport: diving penguins, schools of jellyfish, water thick with plankton. "I was surprised at how biologically active the cold latitudes were," he remembered. He was seeing what the ocean had looked like before industrial overfishing, pollution, and climate change had wrought irrevocable damage. One of the most abundant life-forms in the Southern Ocean is the Antarctic krill, a shrimplike crustacean that swims in swarms so dense they stain the sea a reddish brown. Krill hold an important middle management position in the Antarctic ecosystem, feeding off the smaller phytoplankton and ice algae and in turn feeding the bigger animals: seals, whales, seabirds, fish, and squid.

But the Southern Ocean is not immune to the forces that have altered the more accessible oceans. Fishing ships that resemble factories more than vessels sail here from as far away as Norway, China, and South Korea to scoop up more than six hundred thousand tons of krill each year, and the crustaceans are then sold into the growing omega-3 vitamin market. In 2006, California banned krill fishing entirely, recognizing that the small animal plays an outsized role in the ecosystem.[6] Less than 1 percent of the Southern Ocean is protected by marine conservation agreements, and most of the waters fall outside national jurisdiction.[7]

When Victor reached the bottom of the trench, he noticed something else: the sediment was dense and coarse and completely unlike the peanut butter–smooth softness of the Atlantic's Puerto Rico Trench. Heather Stewart, the geologist on board, had told him he might find volcanic rocks scattered around the seafloor, too.[8] Very few people on Earth have such insight into the bottom of one seafloor trench, let alone more than two trenches in different oceans. Victor was becoming that rare person who could tell the world, from firsthand experience, how diverse the seafloor truly is.

After two hours roaming around the South Sandwich Trench, Victor

and the submersible reemerged at the surface. The depth reader recorded 7,434.6 meters (24,391 feet). Just as in the Puerto Rico Trench,
the mappers' prediction had been just a few feet off the final number.
"It was beautiful," remembered Bohan with a wide grin.

2.

Though the mappers on board *Pressure Drop* were elated, things were
not going so well for the science team. Chief scientist Alan Jamieson
had sailed to Antarctica before, and the Scot was having a miserable
time. "The weather sucked," he remembered. "It was just spending five
or six weeks getting pummeled every day." Thick wet snow fell on the
science crew as they moved around the ship decks in bulky survival
suits. Whenever *Pressure Drop* rolled in the waves, the scientists clung
to the icy railings lest they pitch overboard into the subzero ocean.

Jamieson had brought with him three metal-framed sampling platforms known as landers. Landers are the Swiss Army knives of deep-sea
science. They can sense and sample almost everything—temperature,
salinity, sediment, depth—and they can capture deep-sea critters.
They come with baited high-definition cameras that draw in and film
sea life, recording the first glimpses of many new species. Compared to
crewed submersibles and ROVs, such as *Hercules* on *Nautilus*, landers
are a budget way to gather as much data as possible in the deep. Jamieson himself had built two of the landers on *Pressure Drop* to the tune of
$100,000. The third lander, purchased with a $100,000 grant, included
a core sampler for retrieving deep-sea sediment.[9]

Deep-sea scientists, hemmed in by the time and money it takes
to go to sea and collect samples, are often forced to ask geographically narrow questions. The Five Deeps Expedition provided a rare
chance to ask big-picture questions about what ties the world's deep-
sea trenches together. Theoretically, each isolated trench could have its
own uniquely evolved set of animals, functioning a little like a reverse
Galápagos Island carved deep into the abyss. Instead, all the deep-sea
critters are similar—why? The expedition geologist, Heather Stewart,
wanted to use sediment samples to piece together the geological past of

the hadal trenches and predict when another landslide or earthquake might occur. There is so much to know about both the geology and the biology of the deep sea that the questions often lead to basic discovery-based research—and scientists need hard data to answer them.

As *Pressure Drop* headed north, over the south 60th latitude line and into the South Atlantic, the weather turned foul. Over the coming days, the ship would work its way north along the South Sandwich Trench, mapping its entire 600-mile length for the first time. Cassie and Bohan stumbled upon discovery after discovery. An uncharted seamount rose up thousands of feet from the seafloor, and as the mountain was coming up on the screen, Bohan's excitement grew. "You're like, 'What do we do with this information? Who do we tell? No one knows about this mountain!'" she said. "It's one of those situations where I imagine that's what astronauts could feel like. You feel very removed from the world."

The science team deployed three landers from the decks of *Pressure Drop* before reaching Meteor Deep, which was thought to be the deepest point of the trench. Jamieson is a pioneer in the technology, and he's deployed hundreds of landers over the years. (His doctoral thesis was titled "Autonomous Lander Technology for Biological Research at Midwater, Abyssal and Hadal Depths.") But so many things had never gone quite so badly for him in a single day. The lander with the sediment sampler malfunctioned and failed to collect a single sample. Another lander never surfaced, and the crew spent three hours scanning the rough seas with binoculars to try to locate it. The third surfaced in pitching waves, and, as the crew scrambled to recover it, the ship's propeller severed the line holding it to the ship. In a second, the lander plopped into the ocean and disappeared, with no way to retrieve it.[10] Jamieson was stunned. In a few hours, he had lost two landers, thousands of dollars of his own equipment, and a vital set of samples from an unmapped hadal trench. Victor proposed naming the new site "Bitter Deep"; Jamieson was not amused.

The next day, there was more bad luck: Jamieson's scheduled submersible dive was canceled due to bad weather. For years, he had deployed landers and operated remotely operated vehicles in the deep sea, pieced together the Hadal ecosystem through video clips and photographs, tubes of sediment samples, and dead animals captured by landers that had not survived the trip to the surface, but he had never personally experienced the Hadal Zone himself. With an estimated five hundred full-time deep-sea biologists in the world and only a handful of deep-diving submersibles, people wait years for a coveted spot on board.[11] Feuds can fester if someone snags a seat before another, more senior candidate does. Jamieson was truly crushed.

The ship continued on to the Meteor Deep, in the middle of the crescent-shaped trench. By now the weather had turned from foul to extremely foul. Mighty gusts raked the ocean, and ten-foot swells roiled the water. The awful weather streaked itself across the EM 124, too. The clean, crisp lines from earlier in the expedition turned to ragged smears. "The reason we had so much trouble is we were pitching so much," explained Cassie, who never once mentioned feeling seasick or even uncomfortable during the leg.

Roll, yaw, pitch: these are the terms for how a vessel moves, and each movement imprinted its effects on the EM 124 a little differently. Roll is when a ship rocks from port to starboard. Sonically, roll creates ripples and bumps at the outer edges of the EM 124's swath. Yaw is when a ship's bow moves from side to side, and its sonar impact is fixable. Pitch is when a ship rocks from bow to stern—and pitching is the worst for mapping. "When a ship pitches, we have a lot of problems because all the bubbles will come down and get shoved under the ship," explained Cassie. The blanket of bubbles underneath the ship blocked the sonar and caused the EM 124 to lose its connection with the seafloor, creating the same types of flyers and smeared data that I had seen on board *Nautilus* off Point Conception.

Despite Victor's successful dive, the morale on *Pressure Drop* had plummeted. Everyone was homesick. No one had seen land in weeks. The ten-foot swells made sleeping almost impossible. The lost equipment, the canceled dives, the miserable weather—all of it had soured the mood on board. The sense of teamwork that Five Deeps had fostered back in Puerto Rico felt far, far away. With the southern hemisphere's winter fast approaching, *Pressure Drop* pulled the plug on the trip and headed north for Cape Town, a two-week journey away. As the ship worked its way toward the northern end of the South Sandwich

Trench, the EM 124 detected another trench branching off from the main stem that had been entirely missed by the satellite-predicted maps. The mappers surveyed the new, unnamed branch and discovered that it—not the Meteor Deep—held the real deepest point in the South Sandwich Trench. In less than a month at sea, the two mappers had upended basic knowledge that had held sway for nearly a century.

"We might have to revise Wikipedia after this," Victor wrote giddily on his blog afterward. As the ship started the long transit to Cape Town, he went to work on the Wikipedia entries for the South Sandwich Trench: "The deepest point below the 60th parallel south, the deepest point in the Southern Ocean, is dubbed by Victor Vescovo as the Factorian Deep, a name that he hopes will become official."[12]

Then, suddenly, a clear patch of weather over the Meteor Deep appeared in the forecast. "[Victor] came into my office one day," remembered Jamieson, and told him that the ship could turn back for a dive in the Meteor Deep. Suddenly Jamieson found himself suiting up for a last-minute dive with Victor. The weather was rough but doable. *Pressure Drop* towed the submersible to the coordinates of the dive site as the two men sat inside, waiting to begin the dive. Then a big wave reared up and sent the submersible colliding with the ship's stern. Jamieson's eyes went wide; Victor assured him that all was well, pointing to the submersible's control board. Every button was lit up green for "good to go." Jamieson and Victor waited a few moments while a launching dinghy motored over to the submersible and inspected the damage from the outside. Then the radio crackled: clear to dive. Victor opened the valves, flooding the submersible's empty ballast tanks and dragging *Limiting Factor* beneath the waves. "When you leave the surface, you leave at three knots, so the sunlight goes in five, four, three, two, one, and it's dark," Jamieson recalled. "It's incredibly graceful."

They began to fall quickly through the ocean and had just reached 1,640 feet deep when the underwater telephone piped up with a directive: abort the dive. After the submersible disappeared, a crew member had noticed that the sub had left a shimmering oil slick on the ocean surface. The crash had severed a cable connecting an underwater camera to the submersible, and the cable was now leaking oil. That would eventually allow seawater to flow into the submersible's junction box and unleash a series of interconnected, expensive problems, and ultimately the submersible wouldn't reach the bottom. Patrick Lahey gave the order to abort. The two men returned to the surface.[13]

Jamieson now felt utterly defeated. He had come within striking distance of seeing the Hadal Zone with his own eyes, only to be turned back by a rogue wave. As the men climbed out of the submersible, disappointment was written all over Jamieson's face. The film team moved in for an interview, annoying him even further.

With the damaged sub strapped snugly back on board, *Pressure Drop* headed for South Africa, in earnest this time. Twelve days later, the team spotted land: a wall of high-rise beach condos ringing the coast of Cape Town. A pod of humpback whales escorted the ship into harbor.[14] Jamieson was still seething over the lost equipment and the squandered dive. He felt there was little science to show for all the months he had spent at sea, apart from a submersible arm sitting at the bottom of the Atlantic Ocean and two landers lost in the Southern Ocean. He started to mull over quitting the expedition. After the ship docked, he flew back to England thinking he would never see the ship or the rest of the Five Deeps team again.

3.

Not long after Five Deeps returned from the Southern Ocean, I caught up with Cassie on Zoom. Over our shared screens, I watched as she rotated, tweaked, and cleaned a 64.5-square-mile map of the Southern Ocean seafloor—an area roughly the same size as Washington, DC. There were thousands of red and orange dots, each one representing roughly one thousand square feet of seafloor. When she pulled back from the swirl, the dots resolved into the clear V-shape of the South Sandwich Trench, a little like an Etch-a-Sketch drawing. "This was a hot, hot mess," she said, pointing to where the pitching ship had smeared soundings well outside the V-shaped trench. Stray dots, known as flyers, were floating at the sides or hovering in the middle of the trench. Although the flyers made no statistical sense, there was always a remote possibility that some might be real. Cassie has spent countless hours sorting soundings, so she worked quickly, deleting stray dots without hesitation. "I might be a bit aggressive in how I clean things out," she admitted, but there is an art to the way a map is built, depending on the person doing the cleaning. "You're never going to get the same map from two different people with the same data set," she explained, removing more dots.

In total, Cassie and Bohan mapped 5,790 square miles in the Southern Ocean—an area roughly half the size of Belgium. Almost all of that terrain was new to science. "We ended up collecting the first complete data set of the South Sandwich Trench," said Cassie. Even though Jamieson detested his time in the Southern Ocean, he had to admit that the new maps were dazzling: "Imagine a blurry old Monet of lily pads, and then Caravaggio comes along and fills it all in."

There was something soothing about watching Cassie piece together the puzzle of the seafloor. It was like watching someone put the world in order again. That was a common reaction among the crew, Cassie said. While she cleaned flyers on board *Pressure Drop*, a biologist or ship's officer might stroll past her desk in the middle of the chaotic dry lab, catch a glimpse of an underwater mountain or canyon coming together on her screen, and become transfixed by the sight of it. It happened so often that she started to call cleaning the map "sonar therapy," partly because of the Zen-like effect it had on passersby and partly because people tended to unburden their emotional state on Cassie while she worked.

According to the memorandum Victor had signed with Seabed 2030, all the maps made on board *Pressure Drop* were headed for the global map. After Cassie finished sprucing up the raw data, she uploaded several terabytes' worth of new terrain onto a hard drive and mailed it to Boulder, Colorado. Inside a gleaming glass-and-brick NOAA building situated in the foothills of the Rockies is the International Hydrographic Organization's Data Centre for Digital Bathymetry (DCDB). This archive, filled with rooms upon rooms of spinning disk drives, holds much of the world's collective knowledge of the seafloor.

Established in 1990 during the transition from paper to digital maps, the DCDB now holds nearly forty compressed terabytes' worth of seafloor soundings. The biggest contributors to DCDB are the nearly fifty ships in the US academic fleet, but more data is flowing in all the time from government, industry, and academic mappers around the world. When Cassie's hard drive arrived at the DCDB, her data would be absorbed into the larger grid. At that point, the mapper behind the map would disappear, and the map would become the property of the world: the first complete map of the seafloor, stripped down to its data points, free and publicly available to all.

It is hard to overstate just how fundamentally different Seabed 2030 will be from the ocean maps that preceded it. For far too long, the modus operandi for ocean mapping was to hoard and hide information, rather than share and collaborate. The Padrón Real, the sixteenth-century equivalent of Seabed 2030, is a good example of that history. This master map of all the new territory charted by the Spanish kingdom probably hung on the wall of the Casa de Contratación de las Indias in Seville, the world's oldest hydrographic office, established in 1503. At the time, Spain and its main maritime rival, Portugal, were sending pilots to explore the New World. By 1500, Christopher Columbus had already made three trips. As soon as a pilot returned to Spain, he was ordered to turn over all his new annotated charts to the pilot major, who ran the casa along with the cosmographer major. Those two men kept Spain's growing knowledge of ocean currents, depths, and coastlines in a coffer under lock and key. State spies and map bootleggers circled the casa, looking to steal classified maps for profit or espionage. After the Venetian pilot Sebastian Cabot attempted to sell Spain's navigational secrets to the English, the king of Spain ordered that all pilots and mates aboard Spanish ships must be Spanish. Portugal took an even more extreme approach. Its pilots kept very few notes or charts on its activities at all. Revealing Portugal's navigational discoveries or expedition plans became an offense punishable by death. The Dutch East India Company, which later overtook the Spanish and Portuguese in maritime expansion, kept its own secret atlas of its route to the Dutch East Indies.

By the eighteenth century, that secretive approach was reaching its nadir. "Most of the 'secrets of the sea' were no longer secret and yet too many ships and valuable cargoes were lost because of inadequate or conflicting information," writes the map historian Lloyd Brown. Most "nations were ready and for the most part willing to co-operate on an international scale."[15] The modern shipping industry we know today is built on cooperation around navigational charts and the sharing of information on the changing conditions at sea.

Though surveying coastlines and oceans is no longer the furtive undertaking it once was, the ocean depths are still shrouded in mystery. In a world where only a quarter or so of the ocean floor is accurately mapped, knowing more about the uncharted terrain than your enemies still carries a military advantage. In 2021, the $3 billion nuclear submarine USS *Connecticut* collided with a seamount somewhere in the South China Sea, a hotly contested waterway in the Pacific. China has spent the last few decades building a case for claiming much of the South China Sea as its own, flouting international laws and customs arising during the millennia that Southeast Asian people have shared these waters. The United States regularly sends naval warships through the South China Sea as a way of asserting its commitment to freedom of navigation on the high seas, but the submarine crash revealed that the US Navy is probably doing a lot more than that under the surface. Defense Department officials declined to specify where exactly the crash had occurred, but, speaking to CNN, the Scripps geophysicist David Sandwell identified twenty-seven uncharted seamounts where the submarine might have crashed by overlaying satellite measurements of the Earth's gravitational field with maps of the South China Sea and then comparing them to GEBCO charts. Those twenty-seven seamounts were missing from the charts.[16] (A scientist acquaintance later told me that the navy had been none too pleased with Sandwell for doing that.)

The prevailing attitude among many nations today is that surveying inside a country's territorial waters is an infringement on its sovereignty. Herein lies a central challenge facing Seabed 2030: How can a complete map of the world be constructed if the world won't cooperate?

4.

On board *Pressure Drop*, Cassie had casually mentioned to Victor that all the new mountains and canyons they were discovering would need names. In the Southern Ocean alone, the team had discovered dozens of new ridges, seamounts, and deeps. Because Victor had funded the expedition, he had earned the right to name them. Victor was delighted with the news. "I had no idea! How do you get to name something?" he wondered aloud. "Do you actually have to touch it? Do you have to map it to a certain resolution? And what is the resolution?" And Cassie said, "Well, yeah. They've never been identified or dived before, so, yes, you'll get to name them."

Victor has a special fondness for naming things. The name *Pressure Drop* and the names of its scientific landers (Flare, Skaff, and Closp) are all drawn from the sci-fi stories of the Scottish writer Iain Banks.[17] He gives all his dogs Russian names (Rasputin, Misha, and Nicholai), while his car names all begin with the letter *G*. He brings a rigor to the task, infusing history, patterns, and inside jokes into each name.

In the Southern Ocean, Victor continued that tradition. He christened many of the newly discovered features after constellations to commemorate the German research vessel *Meteor*, which had discovered the Meteor Deep in the 1920s. But for other features, he deviated from the stars and paid tribute to the story of the Five Deeps. Just as he had promised, he named a deep point of the trench where Jamieson had lost his scientific landers the Bitter Deep. "When he first saw that he was upset, but later he thought it was fine," said Victor.

Very rarely does someone like Victor Vescovo come along: a multimillionaire explorer with top-of-the-line surveying equipment who simply gives away terabytes' worth of high-quality maps. Seabed 2030 can't offer much in return for the donation, but naming rights is one carrot. "It's one of the only incentives for deep-sea mapping right now," said Cassie. As the lead mapper on Five Deeps, it was her job to compile the scientific evidence to support Victor's claim to name. Cassie combed through the seafloor maps and picked out little nuggets of data that identified newly discovered seafloor features. The work was trickier than she had expected. It wasn't always clear how to delineate where one seamount ended and another began. The work felt at times more like art than science.

Along with Alan Jamieson, Cassie packaged each new feature into a separate proposal, complete with accompanying bathymetric maps, descriptions, coordinates, and polygons that pegged the new feature to the seafloor. Then they submitted the paperwork to the official naming board for the seafloor outside national jurisdiction: GEBCO's Sub-Committee on Undersea Feature Names (SCUFN).

Scrawling a name across a chart might sound like the final flourish in assembling a map. In the ocean-mapping world, it's more complicated and politically fraught than that. SCUFN (pronounced "scuff-in") is one of many subcommittees working under GEBCO and shepherding data into Seabed 2030. They all go by guttural alphabet-soup acronyms, such as SCRUM, SCOPE, and TSCOM. Each one is composed of a dozen or more members, most of them scientists, who debate and discuss one narrow aspect of mapmaking, such as technology upgrades or public outreach. There's little money and even less glamour in the Herculean effort. For most, it's a passion project taken on in addition to formal jobs at universities or government hydrographic offices. Though SCUFN's dozen members are also unpaid experts, the subcommittee is far more formal and rigid than the other committees, with a labyrinthine structure and established rules for becoming a member. A little like the UN Security Council, with its five permanent member nations, certain countries have to be included. A Russian member, for instance, has sat on SCUFN for forty years.[18]

"[SCUFN] has a much more formal legal mandate about the naming of seafloor features, and that becomes incredibly political and still is incredibly political to this day," the longtime GEBCO member Robin Falconer told me. "In a way that's different than just the rest of making a seafloor map."

As a rule, physical geographers try to avoid political disputes. They're more interested in the terrain than in the humans who occupy it.[19] In the past, Falconer has warned other scientists that working with SCUFN is a challenge. A volunteer body of ocean experts might not seem like a particularly influential group, but SCUFN occupies a curious place of power. As the Seabed 2030 map grows, political maneuvering is going on behind the scenes to assert national interests on the international seafloor. Whereas Victor has a rather innocent stake in naming the seamounts and deeps he discovered, there are powerful interests at play in naming, and maybe one day claiming, the seafloor.

NAMING AND CLAIMING THE SEAFLOOR

1.

"So, compromise is not my middle name," said a burly, straight-talking New Zealander named Kevin Mackay, whose middle name is actually Arthur. Mackay was addressing SCUFN, trying his level best to convince the eleven other members to approve the seafloor names in the Southern Ocean proposed by Victor Vescovo.

"I'm for accepting all those names. I like the stories behind them," he said of the dozen that had been proposed. He was particularly taken with Bitter Deep. "The guy lost two landers, including one that surfaced underneath the propellers of the ship. Lost all their data. It was not a good day. He was *bitterly* disappointed." Mackay chuckled, leaning back in his chair at the National Institute of Water and Atmospheric Research (NIWA), New Zealand's version of NOAA, in Wellington.

A few SCUFN members were not so impressed with Bitter Deep. One member in Tasmania pointed out that losing expensive equipment at sea is pretty banal in marine science. "Just two weeks ago, we lost a third of a million dollars' worth of equipment in a very *bitter* situation," said Mike Coffin, an oceanographer at the University of Tasmania's Institute for Marine and Antarctic Studies in Hobart. But the more troublesome issue for Coffin was not that the name Bitter Deep lacked originality—although in his opinion it did—but rather that almost all of Vescovo's proposals had failed to follow the SCUFN bible for naming the seafloor, known by the uncharismatic document name of B6.

"As a new member I'm trying to follow [B6], and I take issue," Coffin said. He began to read from the document: seafloor names are supposed to commemorate "ships or other vehicles, expeditions or scientific institutions." A name can honor a famous person, too, but that person should be dead, and he or she should have contributed to ocean science in some way. "Clearly the names we accepted yesterday and the names we might accept today do not follow these guidelines," he said. Two other members on the virtual call nodded in agreement. Making an exception for Vescovo might weaken the rules everyone else was supposed to follow. For the moment, the Sub-Committee on Undersea Feature Names was at an impasse.

I was listening to the exchange over headphones at my home in San Diego in late 2020. It was just a few days after Joe Biden had won the presidential election, and SCUFN was hosting its first virtual meeting in its forty-five-year history. The annual five-day meeting, scheduled to take place that year in St. Petersburg, Russia, had been called off due to the COVID-19 pandemic. Instead, the SCUFN members met virtually over six tightly planned hours spread across two days. They planned to process more than fifty new names for seafloor features, which was fewer than usual to accommodate the new virtual format. At the last in-person session, SCUFN had processed nearly two hundred new names.

There was no gaveling in at the start of SCUFN, but it felt as though there should have been. In what was now the tenth month of pandemic Zoom calls, it was by far the most formal virtual meeting I had ever attended. The dozen members called in from Australia, China, France, Italy, Japan, Kenya, Malaysia, Mexico, New Zealand, Russia, South Korea, and the United States. A dozen or so observers had also gathered to watch the seafloor christened, seamount by canyon by ridge. I was the only journalist among them, the first journalist ever to observe a SCUFN meeting. My request had apparently caused some tension among the members, some of whom felt that the meetings should remain private. In the end, I was allowed to observe under the condition that I not ask questions until the end, and only of members who agreed to be interviewed.

It was a practical impossibility to accommodate all the time zones represented on the call, so the meeting kicked off at 7:00 a.m. Central European Time. Here in San Diego, that meant the meeting began at 10:00 p.m.—late, but at least I didn't have to wake up in the middle of the night. On the East Coast, where local time was pushing past 1:00 a.m., a wan observer sat slumped at her desk. The computer glare reflected in her glasses made it hard to tell whether she was awake. In South Korea, people sat in offices wearing face masks. In Japan, the SCUFN member was sandwiched between two flags hanging on poles. In Vietnam, a representative placed a "Vietnam" placard in front of himself, as though he were attending a meeting of the United Nations. In the United States, things were a little more laissez-faire, with the SCUFN member calling in from a darkened dining room in Springfield, Virginia, a wall of framed children's drawings behind him. "Over," he said after nearly every statement, giving away his military background.

As the meeting got under way, I scooted closer to the screen, straining to hear the voices over the rain violently pounding the roof above my head. After another California summer of record-setting wildfires, the first storm of the winter had arrived. In the next room, my husband snored softly. My dog, curled up nearby, cracked open an eye every so often to peer up at me accusingly: "What are you still doing up?" Tonight, I was staying up to watch how a map of the seafloor comes together, name by name. On day two of the meetings, the proceedings kicked off with Victor's twelve names for new features discovered in the Southern Ocean.

"A point of order: Do we have to have a unanimous decision here?" Kevin Mackay cut in. In Wellington, where Mackay sat, it was now half past seven in the evening. The sun was slowly setting in the office window behind him, plunging a backlit mountain into darkness. The discussion had wandered off track and into a disagreement over the wording of B6, the SCUFN bible. Mike Coffin in Hobart pointed out that the B6 calls the naming rules "principles," and shouldn't a principle be sacrosanct? Coffin and Mackay went back and forth a few times on the proper definition of *principle*. The chair of SCUFN, a patient professor of geophysics from South Korea named Hyun-Chul Han, clarified that he wanted the debate on Bitter Deep to play out.

Then the group could apply their decision to the eleven other names Victor had proposed and move the meeting along. The mounting backlog of new seafloor names awaited. In 2013, SCUFN had processed 53 new proposals; in 2018, the number reached 281—a quintupling over the intervening years. At SCUFN's next virtual meeting, slated for early 2021, Cassie Bongiovanni planned to submit ninety new proposals from Victor Vescovo alone. Vietnam and China are submitting record numbers of proposals each year. The Philippines and Malaysia are catching up as well.

A half hour later, Bitter Deep was approved and the subcommittee came to a decision on all the names submitted by Victor Vescovo. Factorian Deep: accepted. Frozen Ridge: amended. The feature is technically a hill, not a ridge, and so it became the far less impressive Frozen Hill. Onward. Hydris Deep: amended. Triton Deep: pending, because a feature cannot be named after a commercial entity, in this case Triton Submarines. On and on it went into the night—or day, depending on where the members were.

All the SCUFN members work for institutions that have some stake in naming the seafloor. Before each meeting, the members marked the proposals according to a traffic light system: red (rejected), green (approved), or yellow (amended). The meetings are long, formal, and a little quirky. Being a SCUFN member means that you have to care intensely about making a map of a place that few humans will ever see and where no one will ever live.

Kevin Mackay is definitely one of those people. For the last two decades, he's been involved with New Zealand's geographic gazetteer.* There are only a few national gazetteers in the world and fewer still that are active, but New Zealand is *very* active in naming the seafloor, according to Mackay. This is thanks to the Treaty of Waitangi, which the British Crown signed with the Māori in 1840, enshrining Māori rights

---

* A gazetteer, for those who don't know—and I definitely did not before writing this book—is a directory of geographical place-names. A gazetteer might also include information on the name's history or origin, a notation about its official or unofficial status, and a description of the place. A gazetteer is often used alongside a map to glean more information about a place.

and language into law[1] and binding New Zealand to co-govern and consult with Māori tribes on naming new land. That means, in Mackay's words, that "wherever New Zealand goes to, Māori go as well." Today, there are more than two hundred seafloor names from New Zealand (Aotearoa) in SCUFN's gazetteer with names such as Kūmara Hill (*kūmara* being the Māori word for "sweet potato").

Mackay became involved with SCUFN after he noticed that the subcommittee was naming features inside New Zealand's Exclusive Economic Zone without consulting the country and, by extension, the Māori people, whose culture has incredibly deep ties to the ocean. At first, he and his team at NIWA thought they would just ignore the SCUFN names. New Zealand could print its own maps with its own names the way most countries have done for centuries. That was easier to do before the internet. Now that the world has digitized many of its seafloor charts, SCUFN's internationally accepted names have more power over smaller regional efforts. Mackay realized that if New Zealand wanted wider recognition for its Māori naming practices, he would have to join SCUFN and gain traction from the inside. In 2018, he became a member.

Not too long ago, chaos reigned in naming the seafloor, and some might argue that it still does today. The marine geologist Heather Stewart remembers that when she started out at the British Geological Survey in the early 2000s, it was still common practice for scientists to name newly discovered features whatever they liked. A scientist would write up a paper describing a new seamount or valley, name it, move on. A few years later, another scientist would "discover" the same feature and give it a new name in another scientific paper, and the cycle would repeat ad infinitum. "There was this guy I know," Stewart said, "and he was desperate to get this little volcanic edifice offshore named Hitchen's Knob."

Slipping naughty names into the seafloor map was the least of their problems. There was a dearth of standardized definitions for describing the seafloor. On land there is some consensus about what makes a valley a valley or a canyon a canyon. We can all see the canyon or valley with our own eyes, and it's far easier to measure and define them above land. Underwater, an equivalent view is more challenging to come by.

With the improvements in sonar technology, a clearer view has only started to emerge over the last sixty years or so.

The earliest way to measure the depth of seafloor was with either a sounding pole or a sounding line, which was limited to measuring a single point directly beneath a surveying ship. Typically, an early surveyor would lower a weighted line into the sea and, when the line hit pay dirt, take a depth reading. When he discovered what seemed like an especially deep point, he might dub it the Challenger Deep or the Tonga Deep even though he had only a glimpse of one tiny bit of seafloor at a time. As the technology improved, multibeam sonar granted mappers a more sweeping perspective of the seafloor. Rather than imagining what might exist between all the little spot soundings, as Marie Tharp had had to do when she drafted her maps in the 1950s and '60s, ocean mappers could see, or rather sound, an entire underwater territory and reconstruct the seascape in three dimensions.

In many cases, the "deeps" turned out not to be the deepest points at all. "There's connotations as soon as you say something is a 'deep'; that implies it's an endpoint, that it's the deepest point of this region," explained Stewart. "So you end up with all these little deeps that don't actually mean anything." One challenge of Cassie Bongiovanni's job on *Pressure Drop* was determining which deeps were truly the deepest in an entire ocean and which had only *seemed* like the deepest back when there was no better way to survey the seafloor.

SCUFN was founded as a way to bring some order to the mess that had been made of seafloor names. Several marine cartography panels tried to take up the problem, but chasing down all the stakeholders took time. The world's international waters are plied by scientists, surveyors, sailors, fishers, and captains from every nation on Earth. Going to sea for weeks or months at a time makes them hard to reach or track down. And once they do make contact, those stakeholders might care deeply about one specific facet of naming the ocean in one specific region of the world, say navigation in the Strait of Malacca or ice coverage off the west coast of Antarctica. They might have zero interest in knowing anything else about the rest of the global map.

Some nations forged ahead with their own naming boards, such as the Advisory Committee on Undersea Features (ACUF) in the United

States and the New Zealand Geographic Board (NZGB). But the larger problem persisted in international waters. A collective international body was needed to agree on a set of standard rules and definitions. As the debate between Mike Coffin and Kevin Mackay made clear, there is still plenty of room for interpretation in the SCUFN rules for naming the seafloor.

Why do we name the seafloor at all? It certainly doesn't make any difference to the animals that live there. "It's quite a human thing to name stuff," explained Heather Stewart, "Whether it's a bit of your garden or whatever, humans have just got a thing about naming stuff." Parents put tremendous thought into naming a newborn, perhaps to commemorate a favorite grandmother or uncle and cast an invisible tether back through the generations. Names are also practical and necessary for a verbal species like ours. Without them, we sink into hopelessly overdrawn descriptions of people, place-names, and objects.

But when it comes to the extreme terrains on Earth and in outer space, why do we need names at all? Scientists usually lead the charge because they need names to categorize and define the natural world. SCUFN's equivalent body in outer space is the International Astronomical Union, which runs a working group of astronomers who oversee naming stars. In Antarctica, the Scientific Committee on Antarctic Research approves names submitted by twenty-two countries with a presence on the frozen continent. Although scientists start with the practical intention of naming places to make them more knowable and navigable, a named place becomes a more accessible and human place, and both good and bad come with that bargain.

In 1975, a precursor of SCUFN met for the first time in Nova Scotia, Canada.[2] There was so much work to do at that first meeting that the group approved only one name, that of a sea canyon off Tierra del Fuego. A decade later, SCUFN emerged with the Scripps Institution geologist Robert Fisher at its head. For the next three decades, Fisher would command SCUFN with an iron hand, setting down rules and guidelines for how the seafloor would be named. SCUFN would be fundamentally different from the other seafloor-naming boards that had come before because it would recruit experts from all over the world to oversee naming roughly 50 percent of the planet's surface.

"One traditional prerequisite of exploration and discovery is the 'right' to name the feature discovered," Fisher wrote in *The History of GEBCO*. "The maps of some remote land areas are replete with surviving personal legacies of nepotism, self-promotion or rough humor. . . . So, in some sectors, is the seafloor."[3]

Here Fisher hinted at the darker side of surveying, specifically that of explorers naming their "discoveries" and, in the process, replacing the names already in use by the Indigenous population. Maps of the Americas are a good example of this, covered in European place-names and scrubbed of Indigenous ones. A colonizing nation could then lean on the new names, as well as geometrically accurate maps, to bolster its territorial claims and subjugate the local population. "As much as guns and warships, maps have been the weapons of imperialism," writes the map historian J. Brian Harley.[4]

That imperial past is still present in international law today, particularly when it comes to territorial disputes. In the South China Sea, the former empires of Great Britain and France could conceivably stake a greater claim over the contested islands than could nearby countries, such as Vietnam and China, whose fishers and sailors have worked the waters for millennia. "Over the centuries, international law has fused the requirements of the dominant states for a system that legitimizes their territorial gains with the legalistic practices of a European civil court," writes the journalist Bill Hayton, an expert on the South China Sea disputes. According to Hayton, this legal system prioritizes written evidence, such as maps and names, over a people's ancestral use or cultural attachments to a region.[5]

As the new head of SCUFN, Bob Fisher wanted to avoid that controversial past and set science as the guiding light for naming the seafloor. He forbade politically motivated names in SCUFN's naming bible. There would be no more naming the seafloor after navy admirals who had nothing to do with ocean exploration. There would be no brand-name canyons or celebrity seamounts, either. There would be guidelines, rules, and principles, as outlined by B6. SCUFN members "are expected to be unbiased, apolitical, free of chauvinism, given to appreciate cleverness or appropriate humor, quick to deplore coarseness, sycophancy or nepotism," Fisher wrote.[6] In short, members are supposed to serve as experts on the seafloor, not state representatives.

But even SCUFN members admit that there is always some nationalism at play when it comes to an international arena such as the seafloor. "We're not state representatives," Mackay agreed, "but the reality is there is always a temptation to put your country's best interests at heart."

In the wee hours of the morning in San Diego, the meeting of the Sub-Committee on Undersea Feature Names ended, with all 53 names decided and another 430 proposals waiting in the queue behind them. In three months, the members would do it all over again, reviewing the next batch of seafloor names. I turned off my computer and dragged myself to bed, body tired, mind racing. The meeting had had a diplomatic, almost staged quality to it. I suspected that things might have gone differently if a journalist hadn't been watching.

Even after I made a mental note to watch the next SCUFN meeting, it took me hours to fall asleep.

2.

At Seabed 2030's Global Center in Southampton, England, Helen Snaith oversees the small team responsible for stitching together the supermap of the world's seafloor. New maps flow in from regional centers all over the world: in Germany, the United States, Sweden, and New Zealand. Each center is responsible for a specific ocean territory. Smaller maps of newly named seafloor features also flow in from SCUFN. As much as Snaith appreciates SCUFN's contributions, incorporating a chart of a small seamount or ridge into a much bigger and mostly blank map is a cartographic nightmare.

"The kind of annoying thing," Snaith said cheerfully, sounding as though it wasn't very annoying at all, is that proposers need only provide maps that support their specific name. "If they're thinking they want to name a seamount or a particular small bay or something like that, we may only be getting like two kilometers squared worth of data. They've got more data around it, but they only have to release that little bit of data." To Snaith, that little oasis of information in an ocean of unknown often indicates that someone out there has more of the map. They just don't want to share it. Yet.

When Seabed 2030 kicked off in 2017, a complete map of the seafloor was estimated to cost between $3 billion and $5 billion. The Nippon Foundation donated $18 million to start Seabed 2030, which went toward setting up a network of regional and global centers and hiring an administrative staff. But obviously $18 million is not $3 billion and will be nowhere near enough funding to finish the job.

Seabed 2030's low-budget approach is to gather maps from research cruises, the commercial maritime industry, and government hydrography offices—a move that has yielded remarkable results so far. In 2019, Seabed 2030 announced that its latest map, which it calls the "global grid," had more than doubled in two years, reaching 15 percent coverage of the ocean floor at a one-thousand-meter resolution.[*7] There is an important caveat, however: very little of the data was technically "new." Most of it had already existed on a hard drive somewhere, and Seabed 2030 had taken the step of tracking down the map and asking for permission to make it public. It seems so obvious, so simple, yet it's extraordinary considering the history of nations hoarding maps. There's an environmental benefit, too. Survey ships run on diesel and produce sound waves that impact marine mammals. Collecting existing maps reduces noise pollution and carbon emissions at sea that sending out an additional survey ship might have produced.

When I spoke with Helen Snaith in late 2020, she estimated that Seabed 2030 was "probably a good year off from actually finding what's out there, let alone actually getting hold of [the data] and getting it into the [maps]."

"We're much farther than that," said Vicki Ferrini a few days later. Ferrini is a geoinformatics researcher at Lamont-Doherty Earth Observatory (LDEO) in Palisades, New York (the name was changed from

---

* In 2022, Seabed 2030 revised how it charted the seafloor. It now aims for sharper resolution in both the shallowest and deepest waters, ranging between 100 and 800 meters.

Lamont Geological Observatory in 1969); she also heads Seabed 2030's Atlantic and Indian Oceans Regional Center. To explain her statement, she outlined a tricky case she had on her hands, tracking down maps from a recent scientific cruise to the Canary Islands. She knew that a research ship had gone to the Canary Islands. She knew that the ship had mapped around the islands. She knew roughly what area it had mapped. But she didn't know who had the data or how to track that person down. Once she did find him or her, there might be more problems in processing the data or getting the data into a format that she could work with. That's pretty typical in the Seabed 2030 world, according to Ferrini. It's a bunch of overworked, underpaid ocean mappers following a trail of bread crumbs around a dark ocean.

Seabed 2030's approach to tracking down stashed-away maps is more than enough work. "It is proving highly elusive," said Snaith of the disparate maps of the seafloor. With more publicity and outreach, she hopes to reverse that relationship in the future. Maybe people will come to Seabed 2030 instead. "If you've got bathymetry data that is sitting on a USB key on your computer, can we have it, *please*?" she pleaded, partly joking, partly serious. "Tell us about it. We'll do something with it."

Snaith would love to see the naming proposals from SCUFN tackle bigger features—the bigger and the more unknown, the better. "If somebody wanted to name parts of the Weddell Sea [near Antarctica] that was like a hundred kilometers squared, that would be great to incorporate, but it tends to be like a seamount or a canyon."

On the global grid, the steep drop-off in data is difficult to ignore. There is much internal agonizing on Snaith's team at the Global Center in Southampton over how to combine the charted and uncharted into one map. Should the beautiful, detailed map of a seamount submitted by SCUFN match the blurry satellite prediction around it? Or vice versa? How should the huge holes in the maps be treated? Sixty years after Marie Tharp used a legend to cover up gaps in the maps, ocean mappers are still grappling with the same problems.

On the Seabed 2030 website, there's an errata page where the public can report errors. "We are kind of reliant on people who are interested in a particular region taking the grid, looking at it and going,

'Well, hang on a minute . . . ,'" said Snaith. "It's usually a scientist who is using the data to study a particular area." She showed me a recent error that a scientist had spotted in the Indian Ocean, where some misplaced coral reefs had turned up on a map near the Maldives. Before launching Seabed 2030, GEBCO received maybe one or two errors a year. Now the mistakes flow in weekly, which Snaith takes as a good sign. "I don't believe it's a reflection that the grid is that much worse than a few years ago," she said. "I think it's a reflection of the fact that so many people are now using the grid."

Seabed 2030 is eager to hear about the mistakes and use the power of the crowd, a little like the way Wikipedia editors use the hive mind to power the world's free online encyclopedia.

Most errors are mundane and can be tracked back to human error, but some intentional errors do crop up now and again, left behind by the Cold War era of cartographic censorship. Today every American with a smartphone uses the Global Positioning System (GPS), but not too long ago, access to GPS was heavily restricted. The US military poured billions of dollars into developing GPS to help guide a ballistic or nuclear missile to a precise location. The first GPS satellites were launched in 1980.[8] Toward the end of the Cold War, civilian scientists began to gain access to the information, too.

Helen Snaith remembered using GPS on a surveying ship in the Southern Ocean in 1995. She was responsible for tracking the ship's position, and she remembered seeing mysterious glitches in the ship's coordinates. "It might suddenly jump you five hundred meters [1,640 feet] that way or five hundred meters that way," she recalled. This is GPS spoofing: the deliberate disruption of satellite signals, an early electronic warfare tactic that the US military deployed during politically tense times. Mappers figured out a way around GPS spoofs by tagging the survey vessel to a nearby station and using it to triangulate their position. But if a mapper wasn't paying close enough attention, the spoofs slipped into maps of the seafloor. Those tiny glitches are still baked into Seabed 2030's global grid today.

Like SCUFN, Seabed 2030 is hell bent on remaining as apolitical as possible. The global grid includes no national borders, and the focus is on surveying international waters—something that will keep Seabed

2030 busy for many years. The group also tries to avoid competition or hierarchies among the regional centers. When I asked Snaith whether the Seabed 2030 Global Center in England served as the hub for the four other regional centers, she corrected me: all the Seabed 2030 centers operate on an equal footing. Kevin Mackay, the New Zealand member of SCUFN who heads Seabed 2030's South and West Pacific Ocean Regional Center, told me much the same. When I asked how much his regional center had contributed to the global map so far, he hesitated to give an exact number. The policy is to avoid competition between nations, he explained, so Seabed 2030 does not release progress reports from its regional centers to the public. Whenever I asked Seabed 2030 mappers outright which countries restricted maps or hid data, they generally avoided naming names. Or they would change the subject by voicing the wistful hope that one day every country would open its vaults and share.

3.

In the coming days, I found out that the SCUFN meeting I had attended would be my last. Moving forward, all meetings would be closed to journalists. The reason, I was told, is that SCUFN was meant to be quick and procedural, and the committee was already scrambling to process 160 proposals. When I promised that I would not slow down the proceedings in any way, the answer was still no. Apparently, the committee had made a unanimous decision to close the meeting to journalists. That was puzzling. The meeting had seemed benign enough. I had followed all the agreed-upon rules. Why wasn't I allowed to observe?

There was still a way to find out what had occurred by consulting the meeting minutes published online afterward. As I read between the lines of the minutes, it became clear to me that the next SCUFN meeting I had missed, in early January 2021, had been not only interesting but also explosive as far as naming the seafloor goes. Vietnam, China, and Malaysia had all submitted proposals for new feature names in the South China Sea—the most hotly contested waterway on the planet.

Vietnam had proposed seventy new feature names; Malaysia had proposed eleven; and China had put forward three.

Historically, Southeast Asian nations have shared the South China Sea. Fishers and sailors from Vietnam, the Philippines, Indonesia, Malaysia, Korea, Brunei, Borneo, and China have all plied the waters, but China was always the big fish in the pond. Over the last decade, China has transformed itself from a regional behemoth into a global superpower with the ability to ban other countries from the water. Chinese coast guard and naval ships have harassed and obstructed oil-survey vessels working within the offshore Exclusive Economic Zone (EEZ) of sovereign nations. Chinese coast guard ships have rammed and sunk Vietnamese fishing vessels found within China's so-called nine-dash line,[9] which is at the heart of the dispute.

China's nine-dash line swoops south from the coast of mainland China along the coasts of Vietnam, Malaysia, Brunei, the Philippines, and Taiwan, enclosing 80 to 90 percent of the sea within a U-shaped area[10] larger than the size of Mexico. In 2016, a tribunal at The Hague found that the line had no legal standing in the United Nations Convention on the Law of the Sea (UNCLOS), the first all-encompassing international treaty for the global ocean[11] and one that China has signed.[12] It also found that China had destroyed the marine environment by dumping sand on reefs and building islands. (Accretion is another time-honored tactic for claiming land by stealing it from the seafloor—the new terra nullius.)

Maritime boundaries are drawn based on the amount of land above the waterline, so the potential for grabbing more ocean territory rests on the distinction between a rock (which can't support human life) and an island (which can).* "As far as maritime resources are concerned, the difference between an island and rock is vast," writes Bill Hayton.

---

* SCUFN does not oversee naming ocean features above the water line or drawing maritime boundaries at sea. The United Nations Commission on the Limits of the Continental Shelf (CLCS) is tasked with implementing the rules laid out in UNCLOS. However, CLCS does not enforce the rules; it makes recommendations to nations on where to draw the outer limit of their continental shelves, which can extend the area in which a country has exclusive rights to seabed resources.

"A rock generates a potential territorial sea of just 452 square nautical miles (π x 12 x 12). An island generates the same territorial waters but also a potential EEZ of at least 125,600 square nautical miles (π x 200 x 200)."[13]

Fisheries are important in the South China Sea, as are the billions of barrels of untapped oil and trillions of cubic feet of natural gas buried beneath the seafloor. But according to a report by the Department of Defense, the dispute is really about China's guaranteeing access to its shipping routes and maintaining its status as a manufacturing superpower.[14] The heavily industrialized Chinese coastline, where sneakers and circuit boards are manufactured and shipped to the world, is surrounded by a string of Southeast Asian island nations, including Indonesia, the Philippines, and, of course, Taiwan, which China already claims as part of its territory.

One of China's softer tactics for exerting control is through seafloor names. In 2020, China announced fifty-five new names for features found along Vietnam's continental shelf.[15] China also trademarked the names and even the shapes of hundreds of islands and reefs throughout the sea—a move that makes no legal sense until you consider the ability of names and maps to bolster territorial claims.[16]

To be fair, China is not the only country playing this game. Japan has spent, according to some accounts, billions of yen building up Okinotorishima, a coral atoll more than a thousand miles from Japan's coast.[17] No one lives on those two islets in the Pacific; the bigger of the two is about the size of a small bedroom.[18] But Japan insists that Okinotorishima is an island and thus entitles the country to a full 200-nautical-mile (230-mile) EEZ and all the marine resources within it.[19] In 2004, the Chinese government said that Okinotorishima was just a "rock" and that it would no longer recognize Japan's expanded maritime border around the islets.*

In general, names and borders in the Asian Pacific are controversial

---

* In response to China's provocative comment, the Nippon Foundation offered to build a $1 million lighthouse on Okinotorishima, if the Japanese government would not do so itself, which would bolster the islet's economic status by guiding ships to shore.

because there are so many coastal and island nations within overlapping EEZ claims. There is no internationally agreed-upon name for the Pacific sea sandwiched between Russia, Korea, and Japan. In Korea, it's called the East Sea; in Japan, the Japan Sea.[20] The name South China Sea is internationally recognized, but within the region the name changes depending on the country. The Philippines claims that part of the sea is the West Philippine Sea; Indonesia has named another part the North Natuna Sea. In Vietnam, it's the East Sea.

In SCUFN parlance, the regions in which countries have competing territorial claims are known as "mutual areas of interest." Parts of the Arctic Ocean and the Falkland Islands (Islas Malvinas) fall under the same bloodless diplomatic term, but "SCUFN really hasn't gotten into [these areas] because those places are coastal," and generally fall in disputed territory, explained Kevin Mackay at NIWA. The subcommittee is responsible only for the international seafloor.

At the January 2021 meeting I was not allowed to attend, SCUFN moved almost all the proposed names in the South China Sea into what's called "mutual consultation," a process that neither approves nor rejects the names but encourages the stakeholders to work out a solution among themselves. Mackay told me he had never seen it work since he joined SCUFN in 2018. The country's state departments usually step in to negotiate. According to Mackay and other members, countries do play games with SCUFN by submitting names for contested seafloor that they know SCUFN will not approve. Behind the scenes, countries try to curry favor for their names. Representatives from South China Sea countries have flown to Sweden and New Zealand to approach SCUFN members at work and at home. In 2020, Mike Coffin, the oceanographer in Tasmania, was still new to SCUFN and had never attended any meetings in person, but he had already experienced some of this backdoor wheeling and dealing.

"I have been lobbied by one of those particular [South China Sea] countries and gotten calls from the embassy in Canberra to solicit my support," he said. "I think it's more the developing countries that see this as an opportunity to make their footprints in the ocean where they've never had a voice before. . . . The United States, Germany, Britain, and France have already charted and named much of the seafloor for over a century now. And if it is within a country's EEZ and there are a bunch of features named in it that have nothing to do with the country whose EEZ it is, I can understand [the developing countries'] frustration."

The chair of SCUFN disagreed, however. Hyun-Chul Han, a geophysicist at the Korea Institute of Geoscience and Mineral Resources, pointed out that the two most persistent proposers are China and Japan: one is a developing country, the other developed. Han believes that the growing push to name the seafloor is a sign of which countries are prioritizing science. But of course, science is not immune to politics. Dating back to the days of the British *Challenger* expedition of the 1870s right through to the Cold War's Space Race, science has long been a tool for demonstrating a country's prestige and military superiority to the international community.

As a South Korean, Han was a somewhat controversial pick to chair SCUFN, as his nation has an economic and diplomatic stake in the South China Sea. He won by a single vote. Since becoming chair, he has worked hard to build trust among the members. Even still, he's planning to quit after just one term in what is typically a two-term position. "I don't want to be a chairman anymore. It's too difficult," he said. "I tried to set up [SCUFN] the right way, especially proposals in the [mutual] areas of interest, how SCUFN should handle those things. I'm going to make certain rules and regulations for that and then resign. That's my policy."

As much as Seabed 2030 tries to avoid politics when drafting the first full map of the world's seafloor, the politics keep creeping back in. Back at the Seabed 2030 Global Center in Southampton, Helen Snaith said that the policy for dealing with conflicting maps is to "de-conflict" the data and merge the two sets together to find the most accurate representation. But sooner or later, a time will come when SCUFN will have to make some tough decisions about naming the seafloor.

4.

Gossip drifts on the air-conditioned breeze of conferences. Off-the-record secrets spill out at postpanel dinner parties along with the wine. Even ocean mappers have their share of conspiracy theories. Some countries *may* have more data than they're letting on. Maybe—just *maybe*—the entire seafloor is already mapped, and all the goods are stashed away on a government's hard drive somewhere. Naturally, the wealthier countries of the global North are the prime suspects because they have the ships and money to pull it off. Tim Kearns, the Canadian mapper who runs Map the Gaps, believed that the United Kingdom, the United States, and the Russian Federation are the nations investing the most in mapping the seafloor. Specifically, a British ship, HMS *Scott*, has spent more than three hundred days at sea each year surveying the seafloor since its launch in 1998.[21] Where have the results of more than six thousand days of seafloor surveying gone? According to what Kearns had heard, no one really knows.

Of course, it's difficult to prove that a country is hiding maps. Even leveling the accusation seems a bit ridiculous, because to a certain extent, every country does it. It's also a strategic advantage for a country to hint that it has more intelligence than it actually does. In the ocean-mapping world, where the military objectives of global powers often lurk behind expensive offshore operations, it's best to question every story you hear. Like when Bell Laboratories built a telegraph cable across the North Atlantic—and a secret enemy submarine surveillance cable for the US Navy—back in Bruce Heezen and Marie Tharp's day. Or the military funding of Bob Ballard's search for the *Titanic*—as long as he found missing nuclear submarines first. Or Howard Hughes's *Glomar Explorer* expedition to mine the Pacific seafloor, which was actually a covert operation to investigate Russian nuclear submarines. There are countless stories like these, declassified years after the fact. So some credence can be given to the theories, partly because they're born in the left-sided brains of skeptical scientists and partly because it's happened before.

In the 1980s, William Haxby, a researcher at Lamont-Doherty in Palisades, New York, created the first satellite-predicted maps of the seafloor using declassified data from the NASA Seasat satellite, which was launched in 1978 and failed three months later.[22] In 1999, David Sandwell of the Scripps Institution and Walter Smith of NOAA refined Haxby's techniques by taking declassified data from the Geosat satellite. Launched by the navy in 1985, Geosat measured the altitude of the sea surface and Earth's gravity field—measurements that had the clear militaristic aim of keeping submarines and ballistic missiles on track. Piecing the declassified data together, Smith and Sandwell showed that the US Navy had mapped the global seafloor, counterintuitively, from the sky. All it needed was a few civilian scientists to dig a little deeper into the data.[23] "Smith and Sandwell ended up finding out how to do satellite-derived [seafloor maps] from satellite data, but that was never the point of sending a satellite up there," said Cassie Bongiovanni. "That was a couple of guys who just figured out what to do with this random data no one was using."

When I spoke to David Sandwell in 2020, he saw himself as something of an outsider in the Seabed 2030 world.[*] Unlike the more cautious Seabed 2030 mappers I interviewed, Sandwell was blunt about which countries are hiding seafloor maps. "It sounds like we share data, but we don't," he said of the United States. "There's about a hundred ship years of military data that's still classified from the mapping during the Cold War. There was a huge mapping effort in the North Pacific and the North Atlantic; they mapped almost a hundred percent. But all that data is still locked up in the vault somewhere."

This policy isn't unique to the United States, added Sandwell. Nearly every country with the means to map has a secret stash. "It used to be the Japanese [who hid maps]," he said, "but they opened up everything a few years ago, so they're really good now. And the UK has been pretty bad, too. They don't share all the Antarctic data still. France is bad. They have some really giant maps in the Indian Ocean around Réunion Island, and you can sort of see where they went, but you can't get any of the data."[†]

At that point, I was mulling over filing a few freedom-of-information requests to see the secret maps that Sandwell spoke of—even though

---

[*] In 2021, the Scripps Institution of Oceanography signed a memorandum of understanding with Seabed 2030. David Sandwell is now working more closely with Seabed 2030 to share and compile bathymetric maps.

[†] Throughout 2020 and 2021, France reversed course and shared maps from the Indian Ocean.

I would never get them until they were declassified decades from now. But not long after, I interviewed John Hall, an American marine geophysicist who cast the race to map the seafloor in a whole new light. During the freewheeling hours I spent talking to Hall, he managed to slam pretty much every entity involved in ocean mapping today. From SCUFN (which had recently rejected his proposal for a seafloor name) to the International Hydrographic Organization ("That's where admirals go to retire"), Hall felt that no one was doing enough to finish Seabed 2030 on time. Now retired from the Geological Survey of Israel, Hall spent thirteen hours every day working on seafloor maps from his house in Israel. "I'll be ninety in 2030," he said, and he was unsure he would see Seabed 2030 complete its mission before he died.

Along with all his time, Hall had sunk a significant sum of his own money into finishing the map. According to his estimates, he had donated a third of a million dollars to GEBCO; over a million British pounds had gone toward developing and operating a hovercraft for mapping around the polar ice cover in the Arctic and Southern Oceans. Despite all the time and money spent, the pace continued to be glacial, exasperating Hall to no end. His pet project is a stretch of Arctic seafloor that could be the site of a possible asteroid collision several million years ago,[24] and Hall has spent years coaxing Arctic maps out of various sources. I thought he might have some juicy stories to share, but whenever I pressed for details, he waved me off the espionage angle. "Ninety-nine percent of the seafloor has absolutely nothing secretive going on except a hell of a lot of biological lovemaking," he said. At that point, Hall was clearly exasperated. Okay, yes, there is some skullduggery going on, Hall said, relenting, and he told me a few stories about declassified Russian maps falling into his hands after the collapse of the Soviet Union and then another story about a British mapper, working in Saudi Arabia, who had told Hall that the king would cut off the head of any mapper carrying around a map without good reason.

But all that political intrigue was beside the point, he insisted. "You know one of the reasons American submarines, much of the time, would not say where they were? They didn't *know* where the hell they were" before GPS, he said. In Hall's strong opinion, secret maps were just a distraction, cloak-and-dagger stories that masked a more obvious challenge to Seabed 2030: the sheer magnitude of the ocean itself.

5.

When the ocean mappers behind Seabed 2030 first hatched their plan, they imagined every country sending forth ships to tackle the last great frontier together. They chose the year 2030 as an ambitious but still achievable target based on the number of dedicated survey ships available at the time. But there was always another idea simmering in the background: crowdsourcing.

Almost every vessel at sea, from aircraft carriers to fishing dinghies, has sonar equipment installed for sounding the seafloor. "By utilising assets already at sea, these craft can effectively become an international fleet of research vessels, collecting and donating bathymetric data to Seabed 2030 for inclusion in the global grid," read an early promotional leaflet for Seabed 2030. Even a basic fish-finding sonar can collect soundings that are more accurate than satellite predictions.

A survey ship costs upward of $50,000 a day, so poor developing nations, without fancy survey vessels, have very few options to participate in Seabed 2030. Tragically, those countries would benefit the most from better maps and have the most to lose from poor ones. As seas rise and coastlines change, the small island states of the Pacific, such as the Republic of Palau, the Republic of Kiribati, and the Republic of the Marshall Islands, are all grappling with a dire need for new and improved charts. Crowdsourcing offers a way into the mapping world without the bigger budget of a developed country. A smaller local survey boat can reach shallow, hard-to-reach coastlines regularly where a larger survey vessel might get snagged or return only occasionally.

Jennifer Jencks felt excited by the potential to crowdsource the seafloor. Jencks, who oversees the NOAA data center in Boulder, Colorado, where all of Seabed 2030's maps are held, also chairs the Crowdsourced Bathymetry Working Group at the International Hydrographic

Organization (IHO). Over the years, she's attended countless ocean-mapping talks and workshops, and she tends to see the same faces over and over again. Not that she's complaining. She loves the small, passionate group of people who care deeply about charting the seafloor. But it was clear that their numbers weren't growing, and neither was the map. To Jencks, those problems seemed interconnected.

"Crowdsourced bathymetry came about a few years ago, when IHO was saying, 'At this rate we're never going to map the whole darn ocean. We really need to start looking outside the box,'" she remembered. As her working group started to spearhead crowdsourcing projects, she began to notice new faces at ocean-mapping events. When we spoke, she had just wrapped up a Caribbean mapping workshop during which Costa Rica had agreed to allow crowdsourcing in its waters under national jurisdiction. "The folks who are coming to the table is so much broader now," she said.

Jencks mentioned a few crowdsourcing projects getting off the ground in Palau, South Africa, and Greenland. One effort was already under way in Australia, where a marine geologist, Robin Beaman, had installed a data-logging system on a dozen or so fishing and tourism boats. Big gaps remain in the maps of the Great Barrier Reef, particularly in the deeper lagoons, channels and inter-reef areas where modern survey ships have trouble reaching.

"Here in the Great Barrier Reef, there are maybe thousands of people with boats, and every one of them has an echo sounder," Beaman told me when I called him up. "Most people just record a waypoint for their favorite fishing spot. They'll put the right point in, drive to it, catch a fish, come home. They don't collect any of the digital data that's coming through that system." With Beaman's data loggers installed, tourist and fishing boats crisscrossing the Great Barrier Reef were filling in blank spots on the map. Whenever he collected a new USB stick, stocked with new maps from a volunteer boat, Beaman said, it was like Christmas; he never knew what he was going to unwrap.

Another project was also starting up in the Canadian Arctic. "There's a big-time increase in [shipping] traffic happening up there," Jencks told me, as much as 25 percent over the last six years, "and barely any of it is mapped."[25] Just like the Southern Ocean, the Arctic

Ocean is big, wild, and uncharted. And much like the small island nations of the Pacific, which are feeling the effects of climate change and rising sea levels first, the Indigenous communities of the Arctic are struggling to navigate the changing environment. After a series of boating accidents in Arviat, an Inuit hamlet in the Canadian territory of Nunavut, the community decided to do something about it. The hunters there couldn't wait any longer for government or industry to come map the water for them. They set out to do it themselves.

CHAPTER 7

CROWDSOURCING A MAP OF THE ARCTIC

Civilization is like a thin layer of ice upon a deep ocean of chaos.

—Werner Herzog

1.

I caught my first glimpse of Arviat, an Inuit hamlet along the western shores of Canada's Hudson Bay, from the rounded window of a twin-propeller plane. A half dozen rows of houses lined up along a shallow inlet that popped a bright Caribbean blue. At the water's edge, the pale green tundra slipped into the ocean gradually and, without any fanfare, land became sea. I had come to see the vanguard in crowdsourcing ocean maps, and this had led me to uncharted waters overlooked by government and industry alike.

The Arctic Ocean is the smallest and shallowest of all five oceans.*[1] I arrived in late summer, the height of beluga-hunting season, when the white whales swim into the bay and everyone gives chase: the killer whales, the polar bears, the Inuit hunters. Summer is also the season

---

* The Arctic is the only ocean without a hadal trench. Its deepest point, the Molloy Hole, at a mere 3.5 miles deep, doesn't qualify.

when cargo ships deliver big items, like ATVs, construction machinery, and building material. The cargo ships never enter Arviat's bay, though; it's too shallow and getting shallower all the time. Also, the charts are old and outdated. Instead, the cargo ships anchor far offshore, sometimes as far as two miles away, in the deeper water of Hudson Bay. A year's worth of goods is loaded onto a flat-bottomed barge and ferried to land over a series of days.

This is how the global shipping industry navigates in the remote corners of the ocean: it stays away. It sticks to well-charted coastlines and transoceanic shipping routes. Europe's largest seaport,[2] in the Dutch city of Rotterdam, is mapped afresh each and every day to capture the changing bottom of the Maas River.[3] A team of surveyors sticks to a similar schedule in the Thames River, which flows into the Port of London.[4] Employees at Boston Harbor create a new map each month. Meticulous attention is paid to charting well-trafficked routes such as these, but with only around a quarter of the seafloor mapped today, well-charted waters are the exception, not the rule. The vast majority of the ocean sees very little shipping traffic at all. Veer off the shipping routes, and a ship quickly finds itself in darkness, cartographically speaking, unknown terrain passing beneath its hull.

There are vast stretches at the polar margins of the planet where no survey ship has ever gone, largely due to the sea ice that once locked up the poles year-round. In the Canadian Arctic, where Arviat is found, just 15 percent of the charts meet international standards,[5] while only 2.5 percent of the US Arctic is mapped by modern methods and equipment. In places where soundings do exist, the measurements might descend from the nineteenth century, when lead lines were dropped over the wooden gunwale of a ship searching for the legendary Northwest Passage. A modern cargo ship threading its way through the uncharted straits and inlets of the Arctic can find itself in roughly the same position as a European explorer setting out across the ocean in the fifteenth or sixteenth century.[6] At the extreme latitudes of the world, the charts are unreliable at best, blank at worst. The locals often rely on memory and traditional knowledge more than they do on maps.

The Arctic is one of the fastest-warming places on the planet, with temperatures rising three to four times as fast as the global average. The dream of finding the Northwest Passage and a faster shipping route between Europe and Asia is finally coming true—at great environmental cost. Each year, the planet's frozen forehead melts just a little bit, raising sea levels and redefining borders; the permafrost thaws, releasing more greenhouse gases into the atmosphere;[7] and more and more ships try their luck at threading through the Northwest Passage. Over the last decades, the number of ships and tourists, the areas visited,[8] and the lengths of the crossings[9] have all been expanding. The year 2012 was a watershed time, with thirty ships making it through the Northwest Passage. (Prior to that, between two to seven ships crossed every year.)[10] By 2040, the Arctic is expected to see a mostly ice-free summer, which will effectively open this shallow, uncharted ocean to commercial transit and development as never before. The ships are coming regardless of whether the Arctic Ocean is mapped or not.

One of the worst environmental disasters in history happened in this part of the world in 1989 when the supertanker *Exxon Valdez* slammed into a reef off the coast of Alaska. Eleven million gallons of crude oil spilled into the pristine Prince William Sound, covering seabirds in toxic sludge and bankrupting the local fishery. The official cause of the accident was overwork and human fatigue, but what was left out of the report is how poorly charted the Arctic Ocean is.[11] Less than a century earlier, a steamship crashed into the same reef and sat there abandoned for an entire decade, serving as a reminder to boaters to keep well away.[12] In 2009, a safety tugboat, deployed precisely to avert another disaster in Prince William Sound, crashed into Bligh Reef and left a three-mile-long diesel oil slick shimmering on the water's surface.

The shipping and cruising industries need better maps of the Arctic Ocean, but so do the people who live here year-round in communities such as Arviat. Life revolves around the water, with Arviat's bay serving as the entrance, the exit, and the lifeblood of the community. In the past, Inuit elders served as repositories of traditional knowledge, called Inuit Qaujimajatuqangit, or IQ, on navigating the environment. A three-day whiteout blizzard has become a thing of the past in Arviat. The bay is freezing later and thawing earlier.[13] Some animals are flourishing, others floundering. When elders share their knowledge, they warn that their IQ might be out of date and untrustworthy because climate change has transformed the world so quickly. Back-country injuries and rescues in the Canadian Arctic territory of Nunavut have doubled over the last decade. Emergencies tend to happen on warm winter days when the hunting is good but conditions on the ice are treacherous.[14] Along with the environmental changes, an obscure geological phenomenon is lifting the tundra and the seafloor. The changes are most apparent in Arviat's bay, where islands and shoals poke above the surface that never used to be there. The land is coming back, as some people say, and the shallowing water has caused an accumulating series of accidents over the years. There are no roads connecting Nunavut's twenty-five communities, so apart from the small airport, visitors reach Arviat by taking a small boat from one of the two closest towns: Rankin Inlet, Nunavut, to the north or Churchill, Manitoba, to the south.

After our plane came to a stop, the fifty passengers on board filed off one by one and walked across the dusty tarmac to the small airport. Almost everyone was Inuit and carrying a backpack stuffed with provisions from the south: cans of Diet Coke, fishing rods, and toiletries. A cluster of children watched and waited at the airport building's picture window, trembling with excitement as the newcomers drew closer to the door. Stepping inside the airport felt like a homecoming: an explosion of noise and energy in the small room packed tight with young families, running children, elders, barking dogs, and piles of luggage.

I set my bags down and turned on my cell phone. My location was a blue dot hovering on the blank grid of Google Maps. "No service," the phone said helpfully. "No internet connection." Before arriving, I had asked Julien Desrochers, the Quebecois ocean mapper leading the mapping sessions in Arviat, how to find him in case I had no cell or internet service when I arrived. For a mapper, he was very blasé about giving his exact location. "Just ask someone where Shirley's house is," he said. Shirley was Shirley Tagalik, a community leader who had hired Desrochers to teach local hunters how to map the changing coastlines.

Pretty soon, I was wandering down the main drag of Arviat, looking for someone who might know where Shirley Tagalik's house was. I knew that her house was somewhere along the bay, but that was about it. The gravel roads were nearly deserted. Now and then an ATV sped past. Then I saw a large group of kids walk by. They looked to be about twelve or thirteen years old, all dressed in athleisure sneakers and sweats, black and white with a fluorescent pop of color. They strutted down the center of the gravel road, one of them holding up a phone blasting tinny music into the crisp Arctic air.

Arviat is a young town with an average age of twenty-five and nearly 40 percent of the population under the age of fourteen.[15] Here, the youths serve as the mapmakers and the inspiration for charting the coastlines. New technology, including cell phones and GPS, has brought conveniences to the north, along with a false sense of security. Of course, distracted teens are not a problem only in Arviat. Human society writ large is suffering from a global pandemic of distraction. But on the Arctic tundra around Arviat, something as simple as forgetting to charge a satellite phone could be fatal. Step outside your front door, and the wilderness begins. A polar bear might be foraging in your garbage bins. An ATV might blast past on the dark road. (Just a few days after I left, a drunk driver struck and killed a teenager in Rankin Inlet, the next town over.)[16] Venture far outside the hamlet's grid of streets, and you could get lost on the flat tundra without any obvious land markers. During the winter, which lasts most of the year, the low land is blanketed in white snow, becoming even harder to navigate.

"Observation is a skill that is being lost, and it's a skill that you need to survive out here," Kukik Baker told me. Baker and her mother, Shirley Tagalik, along with others, help operate the Aqqiumavvik Society, a nonprofit organization in Arviat. Wary of outside experts who drop in with new ideas and then disappear with their discoveries, the Aqqiumavvik Society takes a local approach. "Anything that we do comes from issues that are raised by the community," Shirley explained to me. Earlier generations honed their powers of observation and self-reliance in small nomadic Inuit groups. The youths today are faced with navigating an Arctic environment upended by global forces: warming climate, new technology, and rising ship traffic. The Aqqiumavvik Society is trying to adapt to a changing world by marrying traditional knowledge with innovative tools, such as ocean mapping.

For a small hamlet with less than three thousand residents,[17] Arviat has an unfair share of problems: rampant food insecurity,[18] grinding poverty,[19] a chronic housing shortage where multiple generations live in run-down matchbox homes built by the government sixty years ago.[20] Not unrelated is the epidemic of suicides that ripple through Nunavut's communities with devastating regularity. People under thirty are hardest hit, taking their lives most often.[21] If Nunavut were its own country, it would have the highest suicide rate in the world.[22]

The Aqqiumavvik Society has developed a range of programs, funded by federal grants, to help instill a sense of purpose and identity rooted in Inuit teachings. It has built greenhouses for growing local fruits and vegetables, which are expensive to ship from the south. It offers classes in cooking country food, hunted and gathered from the land, with recipes such as marinated maktaaq (whale skin) salad and tuktu (caribou) stir-fry. But the most successful program so far is Ujjiqsuiniq Young Hunters. Children as young as eight head out onto the tundra with experienced hunters and learn Ujjiqsuiniq: how to observe and steward the environment. Since Ujjiqsuiniq Young Hunters began in 2012, not a single participant has committed suicide.

More than 70 percent of the global seafloor is still uncharted, a statistic that includes deep international seafloor far from land, just as many people might expect, but also the remote coastlines of developed countries such as Canada, where Indigenous people live, travel, and work. The Aqqiumavvik Society's latest push is to chart those uncharted coastlines at last. For years, stories circulated in town about accidents involving experienced hunters running aground on shoals and reefs rising out of the water unexpectedly. Sometimes those accidents were fatal.

2.

On an August evening in 2014, three hunters set out in two boats to check their traps in a popular fishing spot 30 miles north of Arviat. When they reached the fishing grounds, known as Sandy Point, one man peeled off from the other two on his own boat, but the two groups stayed in contact via CB radio. When the two men went silent on the radio, the lone man returned to look for them and found their drowned bodies flung clear of their boat, which had collided with a reef or a shoal.[23] Sudden deaths are all the more traumatizing in northern communities such as Arviat, where tragedies play out in real time on the CB radio for the whole hamlet to hear: a man hailing his lost friends on the radio again and again as he searches, the silence dragging on, everyone listening and waiting to hear the news.

Just a year before the two men drowned at Sandy Point, Joe Karetak had a near-death experience on the waters near Arviat. A lifelong hunter, Karetak was born in Arviat more than sixty years ago. He's spent his whole life navigating its waters, frozen or not. On an unseasonably warm January day in 2013, he and his son set out on a wooden two-person dinghy to hunt seals. In the afternoon, the wind picked up rapidly and blew the men off shore. Soon, father and son found themselves stranded and adrift on an ice floe. A helicopter sent to rescue the pair landed on thin ice and then crashed through into the ocean right in front of Karetak. He watched, helpless, as the pilot kicked his way out of the sinking cockpit and swam to the ice floe. The two men then became three, waiting, wet and shivering on the ice, for another helicopter to come. All three men survived, but Karetak, who is now in his sixties, said that his body has never recovered from the long hours he spent waiting on the ice floe in wet clothing. There are certain unpredictable spots around Arviat where he won't fish anymore, places that are so shallow that the Arctic char have stopped coming.[24]

"Some of it is self-inflicted because we hunt whales in the shallow water," he said of the boat accidents around Arviat, "but this summer, I thought I was far enough from the shoreline in low tide and I hit." He managed to salvage his bent propeller shaft, but a broken lower unit—an integral part of a boat's engine—is a common story in the hamlet. A lower unit costs a few thousand dollars to fix, plus the time it takes to receive new parts shipped from the south. The time and money can put a hunter out of commission for months, if not years. If a seasoned hunter such as Joe Karetak can't travel the ocean safely anymore, Inuit skills and hunters are endangered.

Distracted teens, a warming climate, a rising coastline, and unreliable or blank charts: these are good ingredients for disaster. "We decided we wanted to create maps so that people can see where these new shoals and reefs are formed," Kukik Baker told me. Outside the bay, there are barely any coastal charts at all, and the locals share tips and tricks to avoid hazards. The Canadian government last mapped Arviat's bay more than twenty-five years ago, and the water has shallowed tremendously since then. Commercial shipping lanes and fishing areas will always be higher priorities in the federal budget, and there's only so much money to go around. "The government was not interested in mapping our coastal waters," Shirley told me. Undeterred, she and Kukik reached out to a small Quebec company, M2Ocean, which led her to Julien Desrochers, the former chief operating officer of the company. M2Ocean makes an easy-to-operate sonar intended for non-experts to map their coastal waters.

You don't have to be a master mariner to see how the ocean around Arviat has changed. As I was walking the hamlet's gravel roads, still looking for Shirley's house, I saw the waters of the bay peeking between the buildings of Arviat's main drag. The locals warned me not to walk along the shoreline; polar bears hang out there, they said. But a quick look couldn't hurt, right? The water seemed so tantalizingly close to the street. Wouldn't I see a giant white bear coming at me across the flat green tundra? I decided to risk it and walked down the short, rubbly incline to the shore.

Over the last century, the land and seafloor around Arviat have risen by more than three feet—something I could see clearly at the shoreline, where the homes sat a good thirty feet back from the water's edge. That gap is growing from a third to a half inch each year. The last ice-age glaciers in Arviat melted some eight thousand years ago, but the terrain is still feeling the effects of lying under a nearly two-mile-thick blanket of ice.[25] Known as postglacial rebound, land and seabed are springing upward little by little, as your mattress does after you get out of bed. Because these ancient ice sheets were thicker in some spots, thinner in others, the elevation gain is not uniform or predictable. Another wrinkle: the terrain doesn't only rise from glacial rebound; it can fall, too. Chesapeake Bay is predicted to sink as much as a foot and a half over the next century because the ancient glaciers tapered out nearby, producing a bulge in the terrain that is slowly collapsing. This is why geologists use the term *glacial isostatic adjustment* to describe Earth's micromovements caused by megatons of glaciers thousands of years ago.[26]

My grade school geology class never covered postglacial rebound. And it definitely didn't cover any of the other mind-blowing action on the seafloor, either: the spreading midocean ridges discovered by Marie Tharp, the abyssal trenches sucking old seafloor down into the Earth's mantle, and the seamounts rising up from hot spots. Most of us overlook the seafloor's role in shaping the planet because the ocean conceals most of the geologic activity. More than 80 percent of volcanic eruptions happen underwater, but we rarely hear about them unless they cause a tsunami or create a new island. Subtle adjustments, including postglacial rebound, are even more difficult to perceive. But the Inuit, whose traditional way of life depends on observing the environment closely, noticed the changing coastlines long before Europeans arrived.[27] Stories passed down from elders spoke of a time not long ago when the roads I was walking in Arviat were underwater. Within a generation, the Inuit's world had shifted.

Almost all of the twenty-five communities in Nunavut, Arviat included, began as a catchment area for nomadic Inuit groups who moved from or were forced off their ancestral land in the first half of the twentieth century. Three distinct groups ended up in Arviat, one of them being the Ahiarmiut, an inland tribe that had lived around 200 miles west of Hudson Bay and survived on caribou.[28] When the Canadian author Farley Mowat encountered the group in the late 1940s, their numbers had shrunk from hundreds just a few decades earlier to some sixty survivors.[29] Under the guise of saving the tribe, the government had roughly relocated the Ahiarmiut 60 miles away without shelter, supplies, or caribou to hunt. (Rumors circulated that the real cause behind the move had been the staff at a nearby weather station who wanted the tribe removed.) Three months later, after a number of Ahiarmiut became ill and died, the members hiked back to their traditional territory. A few years later, the government moved them again, this time over a hundred miles away, where desperate conditions killed off seven more members.[30] The Canadian government moved many other tribes around the north like this, but the Ahiarmiut's story is an extreme case, and Farley Mowat's best-selling books publicized their brush with extinction. Today, the name is a sort of shorthand for cultural genocide in the North. The surviving Ahiarmiut eventually settled in Arviat and two nearby communities. In 2019, a Canadian government minister made an official apology to the twenty-one survivors at a community hall in Arviat. One survivor, Mary Anowtalik, who had been just a young girl when she had watched government agents plow her family home into the tundra,[31] bent her head into her hands and sobbed.[32]

Today, old and new, past and present swirl in Arviat, like oil in water. There are a Tim Hortons coffee shop and a Kentucky Fried Chicken on the main drag. At the three grocery stores in town, young mothers wheel shopping carts up and down the brightly lit aisles, babies strapped to their backs in traditional Inuit parkas. There are caribou antlers piled on the roofs of homes and caribou pelts strung across porches. Standing down by the water, I saw the evidence of a recent beluga kill scattered around me on the pebbly shores, including a severed tail, its gray skin crosshatched with scars. Nearby was a handmade sled, recycled from the wooden shipping pallets that southern goods arrive on. Known as a *qamutiq* in Arviat,[33] such a sled can be made from all sorts of materials and used to haul a whale carcass up onto land. I would find out later that nearly every beluga caught in Arviat has claw marks raked along its back by polar bears prowling the shores.

Polar bears spend much of the summer swimming, and they can emerge out of the water quickly, like alligators. In 2018, a polar bear killed a young father when he and his daughters went to collect bird eggs on an island just outside Arviat. The man didn't have his gun on hand when the polar bear charged the family; his children narrowly escaped.[34] Learning all that would make me feel very stupid later on for ignoring the local warnings and walking around so casually by the shore, alone and unarmed.

A cold breeze swept across the bay, and I jammed my hands into my pockets, wishing I had brought gloves. The sun would set soon. I needed to find Shirley's house before dark.

When Shirley and Kukik had reached out to the Quebecois mapper Julien Desrochers at M2Ocean, he had already introduced an early ocean-mapping prototype to three Arctic communities. Unfortunately, the tool hadn't caught on as Desrochers had hoped; in one community the survey system had been left abandoned in a shed. Arviat, he thought, might be different. The community had reached out to him, not the other way around. He was keenly aware of Canada's fraught colonial relationship between north and south: the way the south sends experts to "fix" the north, offering aid that the Indigenous community never asked for.[35] But for Shirley and Kukik to ask him to come, that was different. That was a sea change.

I turned back from the water and trudged back up the shore. At a crosswalk near Arviat's main street, an official-looking truck pulled up beside me. I waved, and the man inside rolled down the window. "Do you know where Shirley Tagalik lives?" I asked. He nodded, opened his mouth as though he was about to explain, and then motioned for me to hop in. Now I understood Desrochers's advice. It was far easier to ask for help here than rely on a map on my phone.

3.

The next morning, Desrochers ran a crash course on mapping the seafloor in Shirley's living room. Dozens of variables go into making a proper nautical chart of the seafloor, but to create a basic map of underwater topography—the kind that Seabed 2030 aims to finish by the end of the decade—all that is needed is three points: depth, time, location. The HydroBlock, a stripped-down sonar developed by Desrochers and the team at M2Ocean, is designed to collect these data points, and others, with as little input from the user as possible.

Unlike the wall of multibeam sonar I had confronted on board E/V *Nautilus*, the HydroBlock is a self-contained survey system that lives inside a lightweight hard case smaller than a carry-on bag. "Single-beam [echo sounder] data isn't necessarily worse than or better than the multibeam. It's just that there's less of it, and so it can take longer to map," explained Jamie McMichael-Phillips, the director of Seabed 2030 and a big champion of crowdsourcing. Basic single-beam soundings are still a vast improvement on satellite predictions. "The magic of crowdsourcing is that [the soundings might be] lesser quality but still hugely valuable as far as the [Seabed 2030] grid is concerned," McMichael-Phillips said. "It's better to have observer data than none at all."

Here are a few things I learned about mapping the seafloor at the training session that morning. First: Go out to sea at high tide, when you can survey more of the shallow nooks and crannies that dry up during low tide. Second: Run your survey lines perpendicular, not parallel, to the shore so that the sonar captures a cross section of the descending depth, rather than running along the same depth. Next: Space your survey lines at a distance three times the average depth. (In Arviat, the bay is so shallow that the survey lines are packed tightly together at 30 feet apart.) Last but not least: Do not gun the engine. A speed of 4 or 5 knots, about the pace of a run, is best for surveying.

That last advice was the hardest for Andrew Muckpah to follow. After the training session, he and the four other Aqqiumavvik instructors trooped outside to run survey lines across Arviat's bay in the afternoon. The orange HydroBlock case was winched to the gunwale of the Aqqiumavvik boat with a metal vise, and then two poles were attached: a satellite sensor sticking into the air that triangulates position and time, and another pole sticking down into the water, sending pings to the seafloor and receiving them back, measuring depth. The hard case also contained an inertial measurement unit (IMU) that tracked the HydroBlock's position and smoothed out the boat's pitch, yaw, and roll. Depth, time, location: check, check, check. Muckpah, who goes by the baby name Balum—which means "fat" in Inuktitut—started driving the boat across the bay at an agonizing 5 knots. He *never* used to drive this slowly, he said. Typically, the goal is to reach the fishing grounds as quickly as possible. It took all of his patience to drive in a slow straight line, rather than blast across the bay as usual. Sitting on the dashboard in front of him, a laptop displayed the survey's progress so far.

Among the instructors, Balum had the biggest stake in surveying the seafloor. A few years ago, he had destroyed the lower unit of his boat on the way back from a seal hunt. It had happened in autumn when the bay turns to slush and the seals haul themselves on top of newly forming ice to sun themselves. That's when the hunters strike. Balum had caught five seals that day—not a personal record but still a good day. A strong hunter, raised in a family of hunters, he had long since shed the baby fat of his youth. When he took off his sporty sunglasses, which was rare, he revealed dark tan lines around his eyes—a sign of long hours spent on water, ice, and snow, the bright Arctic sun flaring off every surface. There was something soft about his features, though, and the sun freckles scattered across his cheeks. Perhaps that was why the baby name had stuck and followed him into adulthood.

The day of the accident, as they were working their way back to shore, Balum's friend took the wheel, following the trail they had cut through the slush on their way out. That's a typical way to navigate around Arviat's uncharted bay during the slush of seal-hunting season, but that day, the trick failed. While they had been away hunting, the tide had gone out and the trail through the slush had shifted, something they had realized only when the boat ended up on top of a rock and the two men were forced to hop down into the shallow water to push it off. "We were hip deep, and the water was probably minus twenty [minus four degrees Fahrenheit] or even colder," he remembered.

Balum had no idea when he might afford to pay the $1,500 bill to fix the motor's cracked lower unit.[36] When we met, his girlfriend had just given birth to a baby girl. Since then, they've gotten married and have a baby boy now, too. Scrounging up money to fix his boat seemed even less likely these days.

The governing wisdom is that ocean mappers, trained at elite hydrography schools such as the Center for Coastal and Ocean Mapping at the University of New Hampshire, make the best maps. But "hydrography data does not have to be the best to make decisions. That's a bit of a myth," said Brian Calder, the associate director of CCOM. Calder has written extensively about the uncertainties contained in professionally made maps, and he's experimenting with crowdsourcing himself. Yes, trained hydrographers make good maps, he said, but so can a "superobserver," his term for an untrained crowdsourcer committed to the cause.

Both Calder and Desrochers dislike the crowdsourcing label. It naturally draws comparisons with the world's best-known and most successful crowdsourcing effort: Wikipedia. The online encyclopedia's legion of unpaid writers and editors validate one another's work through sheer critical mass, eventually arriving at an approximate version of the truth. When you read a Wikipedia entry, you're vaguely aware that it isn't the most authoritative take available, but for settling a disagreement with a family member, it's often good enough. Crowdsourcing a seafloor map follows roughly the same principle: a critical mass of volunteers surveys the same seafloor and eventually arrives at a reasonably accurate measurement.

HydroBlock doesn't work that way, explained Desrochers. An expert guides a nonexpert through the process, and the survey system's more precise GPS creates a better map. A typical smartphone's GPS can pinpoint location to within 4.9 meters (16 feet) at best. HydroBlock's GPS pinpoints location to just a few inches—a scarily accurate level of precision.[37] As Balum steered the boat across the bay, Desrochers gave the occasional tip or correction as he surveyed.

"When you're running lines to do a survey, it's not only about going straight, right?" Desrochers yelled over the wind. The bright summer sky of the day before had turned gray, overcast, and windy. The bay was awash in little sheep, said Desrochers, using the rough French translation for *whitecaps*. "It's a body of water," he continued. "It's moving so you have to take the current, the wind, all that stuff, into account and adjust your course."

"How hard is it, Balum?" I asked over the wind. Almost on cue, the boat began to stomp over a set of waves. The laptop slipped off the boat's dash, Desrochers catching it before it fell. When he set it up again, the screen showed that the boat had swerved a few meters off course. Balum wrestled with the wheel a moment.

"It's always different. The wind is going this way, that way," he answered as he got the boat back on track. It required just enough of his attention that he couldn't do anything else, including look at his phone. Meanwhile, the tide was going out. Each time we completed a round trip across the bay, the tide had pulled back a little more, leaving a wet rim in the sand.

According to Desrochers's calculations at the time, it would take two days (fourteen work hours) to survey the entire bay. That was the plan on paper, at least. In reality, things were more complicated. Summer is the time for catching beluga, and a good hunting day tends to trump all the other tasks of the Ujjiqsuiniq Young Hunters. Beluga is an important local food source for Arviat; the meat is shared among the community and preserved for the long winter. The instructors also perform biopsies on the whales, collecting samples for environmental monitoring. The unofficial leader and the only woman in the group at that time, Aupaa Irkok, reckoned that the instructors could find the two days to finish mapping the bay before it froze if they swapped off duties. By next summer, when the ice thawed and boats set out to hunt again, the Aqqiumavvik Society planned to share an updated map with the community. That's a key tenet of Inuit identity. Information has no value unless it is shared, the elders say.[38]

Over the last decade, the prejudice against crowdsourcing ocean maps has started to wane, said Mathieu Rondeau, a colleague of Julien Desrochers who helped develop the HydroBlock. Back in 2010, Rondeau was just starting out in crowdsourcing maps when he gave a talk on the topic at a hydrography conference. Very few people attended, and no one asked him any questions when he finished—the academic kiss of death. Rondeau now works at the Canadian Hydrographic Service (CHS), where he says he's seen a sea change within the community. Although some government surveyors are still resistant to crowdsourcing, Seabed 2030 and other major mapping groups have thrown their weight behind the effort.

Perhaps there's also a grudging acceptance that trained hydrographers cannot finish Seabed 2030 all alone. There's simply too much ocean to map, as John Hall pointed out. The CHS is tasked with mapping nearly 800,000 square miles across the Arctic, only 15 percent of which has been accurately surveyed. "It's very expensive to prepare a survey boat in the south and then get to the north and to have hydrographers on board and to survey," Rondeau explained. It takes days, if not weeks, for an icebreaker to smash its way through last winter's ice. Add in the cost of food, fuel, the staff, ship time, and renting expensive equipment, and the price of an Arctic expedition can easily climb past the million-dollar mark. "I could imagine it could be less expensive to leverage the people who already lived there, to equip them, to train them, and to do the job for us," Rondeau explained.

In the Arctic Ocean, the increased shipping traffic makes new maps all the more pressing. The CHS encourages vessels to follow established shipping routes through the Canadian Arctic, which are better mapped, at about 42 percent. But cruise ship captains don't always listen. In order to give well-heeled passengers the "real" Arctic experience, cruise ships veer off into uncharted fjords and inlets so their clients can see towering steely-blue glaciers, long-toothed walruses, and pods of belugas. In 2010, the cruise ship *Clipper Adventurer* plowed into an underwater cliff off the coast of Nunavut. It took two days for an icebreaker to reach the ship and rescue the passengers on board.[39] A year before that, *Clipper Adventurer* had rescued its sister ship *Ocean Nova* after it had run aground in Antarctica.[40] More and bigger ships are coming to the Arctic. The first mega-cruise ship, *Crystal Serenity*, made it through the Northwest Passage in 2016.[41] In 2020, there were fifty-eight shipping accidents within the Arctic Circle, the highest total over three years.[42]

Back at the CHS, Mathieu Rondeau is trying to convince his superiors of the benefits of crowdsourcing: the rising number of accidents, the staggering amount of territory left to survey, the sheer cost of sending a ship to the north. The typical counterargument is that crowdsourcing could lead to a major shipping accident. A nautical chart is technically a legal document; it is constantly updated to support the international maritime industry and support damage claims. Accidents at sea are astronomically expensive, if not outright ruinous to the company involved and the local economy. When the *Ever Given* container ship blocked the Suez Canal for six days in 2021, it halted nearly $10 billion in trade, and the insurance and legal costs ran into the hundreds of millions of dollars. A crowdsourced map didn't lead to the *Ever Given* blockage, but this is the type of headline accident that hydrographers raise when they warn against nonexperts contributing to maps: Who will be responsible when something goes very, very wrong?

Perhaps the most compelling argument for crowdsourcing maps is not cost or efficiency but the danger of repeating the colonial history of surveying in the Arctic. In the past, governments in the south carved up the North as they saw fit, running borders through traditional territory: the Sámi people, for instance, are scattered across Sweden, Norway, Finland, and Russia;[43] the Gwich'in straddle the US-Canadian border between Alaska and the Yukon and Northwest Territories; and the Inuit live across the United States, Canada, Greenland, and Russia. The names of European explorers are scrawled all over the maps of the Arctic: Foxe Channel, Frobisher Bay, Hudson Bay.[44] Some features are even named after wealthy patrons who never set foot in the Arctic, such as Boothia Peninsula and Felix Harbour, named after the gin magnate Felix Booth, who footed the bill for an 1829 expedition by the Englishman John Ross to discover the Northwest Passage.[45] Indigenous names of trails, routes, and geographic features were left off the map, and with them disappeared a deeper Indigenous knowledge of the territory, one that understood the Arctic outside the narrow European focus on finding the Northwest Passage.[46]

As the polar ice caps melt and the Arctic Ocean becomes more navigable, northern nations are narrowing their sights on a new target: the North Pole. In 2007, a Russian explorer and member of the Russian Duma planted his nation's flag on the seafloor at the North Pole.[47] A Canadian foreign minister at the time compared the flag planting to the land grabbing of European colonizers back in the fifteenth century,[48] but Canada plays similar games, too. The federal government has issued Canadian passports to Santa Claus and Mrs. Claus, living at the North Pole.[49] The Conservative government of former prime minister Stephen Harper spent millions of dollars searching the Arctic for the lost ships of the British explorer Sir John Franklin. When Franklin's flagship, HMS *Erebus*, was discovered in 2014, Harper personally announced the news, proclaiming that the Franklin expedition had "laid the foundations of Canada's Arctic sovereignty" more than two hundred years earlier. The shipwreck of a non-British explorer, Roald Amundsen—the first European to successfully navigate the Northwest Passage—was repatriated back to Norway against the wishes of the community in Nunavut, where the ship had sunk, as well as against those of shipwreck archaeologists.[50]

Compared to the heated disputes over the South China Sea, the geopolitical climate in the Arctic Ocean is cooler—for now. But northern nations are still gearing up for a future without year-round ice. Russia, Canada, and Denmark (via Greenland) are all following international law by submitting applications to the UN Commission on the

Limits of the Continental Shelf (CLCS) to extend their Arctic territory. According to the United Nations Convention of the Law of the Sea, a coastal nation's Exclusive Economic Zone extends 200 nautical miles (230 miles) from its coastline. But if a nation can prove that its continental shelf extends farther, it can claim exclusive economic rights over more seabed. It must provide that proof through mapping the seafloor and collecting detailed bathymetric imagery and sediment samples from the seabed. Such surveys are expensive—Canada has spent more than CAD $117 million on seventeen Arctic mapping expeditions so far—but the marine resources a country stands to gain could be priceless. Australia and New Zealand have made successful claims to the CLCS, Australia adding nearly 1 million square miles in 2008 and New Zealand gaining control over 656,373 square miles of seabed, which is six times the size of its land territory.[51]

There are no international stakes involved in mapping Arviat's bay. A new chart of the bay is a good example of what a map would look like if it were made by locals for locals. Balum, Aupaa, and the other Aqqi-umavvik instructors chose to survey the bay first, knowing just how important the waterway is to the community. "It's very powerful that the Arviat community took the initiative to buy and equip themselves with survey equipment and start to do the job by themselves," said Mathieu Rondeau at the CHS.

After running a few survey lines across the bay, Julien Desrochers and the instructors trooped back to the warmth of Shirley's living room. As we sipped coffee, the instructors pored over the maps they had just made, murmuring to one another in Inuktitut, the language they spoke when outsiders such as Desrochers and I weren't around. On Shirley's TV screen, Desrochers brought up the familiar swirl of soundings that I had seen on Cassie's screen and on board E/V Nautilus. He began to pick through the dots, cleaning out flyers and explaining how to process the data before sending it to him for review. When his explanations grew too technical, the instructors retreated into their phones. Mapping the ocean felt too distant from everyday life in the north.

In Arviat, hunting is cool. Men and women dress in camouflage from head to toe. They carry camo phone cases and bags. A good hunter is revered, as a good provider and Inuk. When someone makes a big kill, she radios ahead, and people come rushing down to the dock to grab a piece of fresh beluga. Hunting and sharing are woven into the community fabric. Occasionally, the instructors complained about the tedium of running survey lines back and forth across the bay. What did it have to do with hunting? Or monitoring the ice conditions? "I just tell them it's something we need to do, that the end product is what we're after," said Kukik. The next spring, when the instructors saw the new maps they had made and shared them with the six hundred or so homes in Arviat,[52] she knew they would be proud.

4.

When Seabed 2030 was launched in 2017, I heard nothing but optimism about finishing the seafloor map by the end of the next decade. The deadline was aggressive, people said, but doable. The experts I interviewed pointed to the tremendous progress Seabed 2030 had made in the early years, leapfrogging from 6 percent completed in 2017 to more than 20 percent in 2021.[53] Major crowdsourcing partnerships were announced, with the offshore survey company Fugro pledging to donate all the maps collected on transits between work sites. An Israeli mapper at the University of New Hampshire set to work recruiting the superwealthy and their superyachts to crowdsource maps on their jaunts around the world.

As the 2020s dawned—along with the COVID-19 pandemic, disrupted supply chains, and the largest land war in Europe since World War II—Seabed 2030's prospects dimmed. When I checked in with Jennifer Jencks about the crowdsourcing projects in Palau and South Africa in 2021, both countries were still waiting on equipment shipments, delayed indefinitely by supply-chain issues. Seabed 2030 wasn't the only grand ambition being derailed. More intangible long-term goals, such as the fight against climate change, were also falling prey to pressing short-term problems.[54]

I heard the optimism start to wane in ocean-mapping circles, too. Tim Kearns, the Canadian mapper at Map the Gaps, told me that he still hadn't seen a viable plan for finishing Seabed 2030 on time. "If we're really going to get serious about ocean mapping by 2030, we need a plan," he said. All the piecemeal projects, such as strapping data loggers onto superyachts and chasing maps locked away in vaults, certainly helped, "but it's not addressing the hundreds of thousands of millions of square kilometers that are never going to get touched by any of those methods." Robin Beaman, the marine geologist leading the crowdsourcing project on Australia's Great Barrier Reef, told me much the same. Crowdsourcing worked well in shallow areas along the continental shelf, but Seabed 2030 needed to enlist deepwater multibeam sonars to tackle the international ocean.

Some mappers started to hint that Seabed 2030 was not just ambitious but probably impossible. "I don't need to think about it a long time to know it's not feasible," Julien Desrochers said on one of our last days surveying Arviat's bay. Look at the maps of the Arctic, he said, or better yet, look at just the Hudson Bay, where Arviat sits on the western shore. Hudson Bay is bigger than Texas, bigger than California; a third of Canadian rivers run downstream to Hudson Bay, along with some major American ones. When you focus on the charts of Hudson Bay, what's been mapped is astonishingly small. There are some modern surveys along the coast and a few safe shipping routes across the bay, but much of its 316 million square miles of seafloor is uncharted. Arviat is only one hamlet on a mostly uncharted bay in a mostly uncharted ocean in a mostly uncharted world. At various points, I asked mappers to show me just how much ocean there was left to chart. Hudson Bay made the magnitude of the uncharted ocean crystal clear. It also made Seabed 2030's deadline look pretty much impossible to me, too.

Crowdsourcing, it seemed, would not be a major force in finishing Seabed 2030 after all, but perhaps that had never been the point. "The community aspect of volunteer bathymetry, I think, has a much better chance of getting people excited about doing something, and in a sense, that may be its value, right?" said Brian Calder at the University of New Hampshire. Watching the hunters chart Arviat's bay, I saw tremendous power come from their mapping traditional waters. Seabed 2030 might not meet its deadline, but a better map of the bay would make all the difference to the people of Arviat—more than Seabed 2030 ever could.

When I finally made it out of Arviat, after thirty-six hours of heavy fog and canceled flights, I stepped into the faster, frenetic world of the south. My phone regained its signal, and a flood of missed texts and calls came pinging through. The blank grid of Google Maps became populated again with well-charted streets. My little blue dot reappeared, tracking my position as I moved around the world. Falling off the map had raised another question for me: If crowdsourcing wasn't the magic bullet for Seabed 2030, what was?

At the launch of Seabed 2030, crowdsourcing was often mentioned in tandem with autonomous mapping, that is, mapping the seafloor using drones or some other remotely operated hardware. As the 2020s wore on, I heard less about people-powered solutions, such as crowdsourcing, and more about big technological breakthroughs. When a big, ambitious goal, such as limiting global warming to no more than 1.5 degrees Celsius—2.7 degrees Fahrenheit—starts to dim, people often turn to technology to make up the downfall.

Denis Hains described himself as an optimist by nature, and he described Seabed 2030's deadline the same way: *optimistic*. As a former hydrographer general and director general of the Canadian Hydrographic Service, he had a unique insight into the country's ocean-mapping capabilities, and he couldn't imagine that Canada would send forth its ships to help Seabed 2030. "In Canada, the only ship that could currently do that is the *Louis S. St-Laurent*, and it is very unlikely that this ship will be allocated to do something other than national priorities unless there was political will to invest the time of the ship internationally," he said. Still, he hoped that Seabed 2030 would succeed. The best way to do that, according to him, was to bring in the drones.

"What I believe in is unmanned surface vessels, the Saildrone type of business model," he said, referring to a California start-up whose ocean drone had successfully crossed half the Pacific Ocean on its own. John Hall, back in Israel, agreed: "The only way we were going to do it is with Saildrone," he said. Saildrone, with its distinctive fluorescent-orange carbon-fiber sail, heralded a new future in ocean mapping: one that required zero people on the ocean at all.

CHAPTER 8

THE ROBOT REVOLUTION AT SEA

1.

What does the future of mapping the seafloor look like? If you're Richard Jenkins, the founder and CEO of Saildrone, it looks like sitting in a sunny conference room overlooking San Francisco Bay, drinking a cup of coffee as you watch an icon on a computer screen move across uncharted seafloor a hundred miles offshore.

That was what I was doing one morning while visiting the start-up's headquarters in Alameda, California. The day before, Saildrone had sent its 72-foot ocean drone, the first of a class called Surveyor, on a testing mission offshore. As Jenkins and I chatted, the Surveyor quietly ran a set of perfectly spaced survey lines back and forth across a stretch of seafloor 100 or so miles off the coast of California.

Occasionally, Jenkins leaned forward in his chair to tap at the little Saildrone icon moving across the Pacific. He checked in on the Surveyor's "engine room," where a dashboard of gauges showed the remaining charge of the onboard solar panels, the diesel reserves, and the wind power and direction. Another page showed a firehose of data streaming in from the Surveyor's sensors, tracking wind speed and direction, as well as wave heights. The Surveyor's more compact cousin, the Explorer, at 23 feet long, is basically a gigantic ocean-measuring machine with twenty-some sensors tracking ocean temperature, salinity, relative humidity, barometric pressure, dissolved oxygen, chlorophyll, and much more. (The company's motto: "Any Sensor, Anytime, Anywhere.")

On another screen, Jenkins scrolled through pictures taken from every angle on board the Surveyor, picture after picture streaming in fresh and refreshed each minute. Over the years that the various drones have spent at sea, Saildrone has amassed tens of millions of pictures in what it calls the largest collection of ocean imagery in the world. Those pictures have been annotated by humans and fed into a patented algorithm that trains the drones to assess the situation based on what they're seeing. Should the drone change course around an incoming cargo ship? Report a suspicious boat to border control? These questions are relayed back to flesh-and-blood human operators on land, who then make the final call. The goal, Jenkins explained, is to decrease the need for humans to go to sea and chart the ocean floor and make ocean mapping faster, cheaper, and more environmentally friendly.

Jenkins toggled over to a real-time view from the top of the Surveyor's sail 50 feet above the water. "The goal here is to give you the same domain awareness you could get from standing on a bridge of a ship but sitting in a chair [on land] drinking coffee," Jenkins said, his English accent so buttery soft that I had to lean forward in my seat to understand him. Finished with the demonstration, he leaned back in his chair. The Surveyor was doing fine, he said, and we picked up our thread of conversation again. Weirdly, the future of ocean mapping felt a lot like checking social media or doing anything else on your phone these days.

When ocean mappers talk about drones or uncrewed surface vessels (USVs) as the only way to finish Seabed 2030 in time, they mean deepwater models such as the Surveyor. There are two smaller nearshore models, but the Surveyor is the first to come mounted with a Kongsberg multibeam sonar that can reach down to nearly 7 miles deep—just beyond the absolute deepest depth of the Mariana trench. Behold, at long last, the tool to finish a complete map of the world's ocean floor!

Over the last few years, Saildrone has run an impressive series of missions—every single one a success. Crossed the Atlantic: check. Circumnavigated Antarctica: check. Sailed the largest USV from California to Hawaii and back: check.[1] The latter was funded in part by Seabed 2030. Larry Mayer, the director of the University of New

Hampshire's Center for Coastal and Ocean Mapping, oversaw the project and planned the Surveyor's route along as much uncharted seafloor as possible. In total, the Surveyor collected data on over 8,000 square miles of uncharted seafloor between San Francisco and Hawaii,[2] which is about double the size of Hawaii's Big Island. Ocean mappers are undeniably smitten with Saildrone.

"If you look at what the Saildrone Surveyors have been doing in the Pacific, the idea that we take an uncrewed vessel and sail it from Alameda all the way to Hawaii and back, even five years ago, everybody would have thought you were crazy," said Brian Calder, the associate director of CCOM. "That's like black magic for mapping."

In the summer of 2021, a smaller Saildrone model sailed through the eye of Hurricane Sam—a category 4 storm off Puerto Rico that ended up becoming the strongest of the year.[3] The footage of that mission went viral, revealing what the eye of a hurricane actually looks like. So what does sailing through a hurricane look like? Very, very messy, it turns out. The footage is such primal chaos that it's hard to tell what's up, what's down, what's in front. The ocean drone rides a wave so tall that the camera appears to be hovering fifty feet above the water, then smashes back down to the surface with terrifying speed and strength. Lightning flashes, wind rakes the waves, the sky is closed off in an oppressive low-hanging wall of white. The field of view is perhaps 25 feet at most. No one had ever seen that before, because until now that was what a dead man saw right before his ship sank.

"That was a good one, driving through a hurricane. It's funny, now I don't get any questions about [the ocean drone's] durability," said Brian Connon, vice president of ocean mapping at Saildrone. Connon is a retired navy captain who spent part of his career working for the Naval Oceanographic Office, the mapping branch of the US Navy. Just as I had with John Hall, I tried to press Connon on the secretive side of ocean mapping. How much seafloor data does the navy have stashed away in its vaults? Has the United States already mapped the entire ocean? "The navy's got a lot of data, but they're not sharing it, for the right reasons," Connon answered, a good sailor to the end.

Saildrone has a couple other ties to the US Navy. In 2022, an Iranian warship briefly hijacked a Saildrone Explorer, deployed by the US Navy in the Persian Gulf, and escalated the already-tense relationship between the two countries. When a US Navy warship and helicopter approached, the Iranian warship cut the drone free and left the area. Saildrone's headquarters are also housed inside the former US Navy air station in Alameda, a port city facing San Francisco. Most of Alameda's downtown streets are filled with abandoned, graffiti-tagged naval hangars, many of them home to the skyrocketing homeless population in the Bay Area. The Saildrone warehouse is the exception. Painted bright orange on one side, the warehouse bustles with energy and activity. As I entered, I was greeted by an iPad on a podium that asked me to sign a nondisclosure agreement swearing not to reveal the patented technology I was about to see on the other side.

Step past that reception and into a huge air hangar, and you're right on the Saildrone assembly line. In front of me were giant rolls wrapped in sheets of carbon fiber and fiberglass. Over there were gigantic ovens where the carbon-fiber sails are baked inside massive sail-shaped molds. Picnic tables were scattered around the hangar, where VPs mixed freely with the general staff. The dress code was de rigueur jeans, beanie, and Patagonia fleece. I snapped a photo of the scene before a member of the marketing team hurried over to stop me. No pictures, she said; they might reveal the patented wing technology that gave Saildrone its edge. The company was nearing its tenth anniversary, but it still had all the excitable energy of a start-up.

"In 2012, it was just me, on the other side of the bay, in a sixty-square-foot room building the first [ocean drone]," Richard Jenkins remembered. He developed the technology for the drone after he had spent ten years smashing the world record for land sailing in 2009. Land sailing, if you don't know—and I didn't—is basically wind-powered racing on land. Sailing over the flat Nevada desert in a green carbon-fiber contraption he had built and honed himself, Jenkins ultimately reached a speed of 126.1 miles per hour.* The world record gave him a certain cool credibility, as well as an origin story fit for a Silicon Valley

---

* In late 2022, the sailing club Emirates Team New Zealand bested Jenkins's record, reaching an ultimate speed of 138.21 miles per hour while cruising across a salt lake in Australia.

entrepreneur. He credits the quixotic quest with providing the necessary research and development to launch Saildrone a few years after.

In the early days of Saildrone, Jenkins hooked up with Eric Schmidt, the founder of Google, and his wife, Wendy, just as they were outfitting their ocean research vessel, *Falkor*. It was the height of the ocean conservation investing phase among the billionaire philanthropist set. Jenkins, however, was unimpressed with the traditional method of doing science at sea. "For the initial ocean drone I was working with Eric and Wendy Schmidt, helping them with their survey research vessel, and it became very apparent to me that what they're doing is just measurement," he said. "There's no point in thirty people on a big ship pushing around one small sensor to map the bottom." He remembered asking the scientists involved, "'Can't you do this with a small robot rather than a hundred-million-dollar ship that burns a lot of fuel?' And the scientists' panel said, 'It's impossible. Never cross an ocean with a small, unmanned vehicle.' That to me is like a red rag to a bull."

Jenkins went on to prove that an uncrewed ocean drone could cross oceans unscathed. The same scientific panel then told him that crossing an ocean was all well and good, but an uncrewed drone could never *measure* anything at sea as well as a scientist could. Jenkins then embarked on a new mission that directly compared the ocean drone's measurements with those taken from a research vessel, proving that it was indeed possible for a drone to sample just as well as a scientist on a ship.[4] "I guess the drive is being driven by the challenge of someone saying it's not possible," said Jenkins. Although his origin story might sound like those of a lot of other entrepreneurs today, he insists that Saildrone is the real deal. Other start-ups prove things with spreadsheets; he proved that a drone could cross oceans, survive hurricanes, and take complicated measurements on the open ocean—and what harsher proving ground is there than that?

In this new world of ocean mapping, people would be superfluous at sea. In all the mapping I had seen so far, from Cassie Bongiovanni surveying the deepest point of all the world's oceans to E/V *Nautilus* charting the seafloor off California to Balum mapping the bottom of Arviat's bay, one factor stayed the same: a human had always been at the helm. Jenkins turned that thinking on its head. People need a ship and a ship needs fuel, as well as all the creature comforts that keep the crew alive and happy on board. The human factor drives up the cost of a survey to tens of thousands of dollars a day; it also limits the time spent at sea because eventually people need to go home. Saildrone estimates that its drone costs a fraction of the price of a traditional survey ship. Its ocean drones are not only more economical but also more environmentally friendly, as they reduce noise pollution, fuel consumption, and carbon emissions.

Saildrone had a higher purpose, Jenkins told me. "The company was founded with the goal of saving the planet," he said before adding one important caveat: "You have to make a company which can tackle the size of the problem. But the only way you can do that is with a commercial company. So while the motivation is planetary health and climate, you have to be a commercial company to be able to make a difference. There's not enough government funds there for you."

When I name-dropped a few of Saildrone's competitors, Jenkins lowered his gaze and smiled, something he did whenever he disapproved. "I don't know anyone who can compete with what we're doing on the screen now," he said softly, gesturing to the Saildrone icon out on the Pacific. But the competition is stiff. Just a few months after we spoke, the Swedish car company Saab found Ernest Shackleton's lost ship *Endurance* off Antarctica with its own autonomous mapping vehicle, Sabertooth. Ocean Infinity, another autonomous ocean-mapping company, is also competing for mapping contracts with government and industry. Not long after the Surveyor returned from Hawaii, Saildrone secured $190 million in venture capital funding[5] and went on a hiring spree. The staff hovered around a hundred people when I visited, and a year later the number had doubled. I could feel the momentum while walking around Saildrone's air hangar. Is the company on the verge of liftoff? Or is it doomed to crash and burn like so many other tech start-ups?

2.

"Did it work?"

Erin Heffron, the ocean mapper on board E/V *Nautilus*, stepped back from the gigantic wall of electronics controlling the EM 302 so-nar. She had just pressed the big "On" button. Nothing had happened. After a day of stormy weather that had blocked all our attempts to survey off the north coast of California, Heffron thought it was time to turn the sonar on again. The two of us stood for a moment watching the electronics until a red light flicked on. In the data lab next door, the computer screens filled with programming commands, each one describing a new instrument booting up in real time. Then Heffron flicked on the mapping software, and big red warning signs flashed across the computers. She hurried from computer to computer, fixing settings and quieting the warning signals.

The future of ocean mapping that Richard Jenkins had laid out at Saildrone sounded impressive, particularly when I thought back to the reality of mapping the seafloor today. Until you've crawled into the belly of a survey ship yourself and seen just how challenging and time-consuming it can be to collect even a small map of seafloor, it's hard to explain why it's taken quite so long to map the world's oceans—and what a quantum leap forward Saildrone might turn out to be. After turning on the back end of the EM 302, Heffron and I headed down into the belly of the ship to make sure all the circuit boards were hum-ming along correctly.

We climbed down a ladder and walked along a narrow hallway lined with the cabins of the mostly Ukrainian crew. A strong fishy odor hung in the hallway. Heffron explained that the Ukrainians liked to catch and salt their own mackerel, which they reeled in during their off-hours. Then we went down another ladder; now we were beneath the water line, and time seemed to leap backward. There was a for-gotten feeling down there, compared to the renovated upstairs. An old rotary phone covered in a thick layer of dust hung on one wall. There were signs posted in German from the E/V *Nautilus*'s past life as a fisheries research vessel of the German Democratic Republic. Then we reached the "sonar room," which was right next to the walk-in freezer with "Proviantkühlanlage" written across it.

Heffron set her heavy-duty industrial flashlight on the floor and used both hands to crank open the steel door. On the other side was a small, cold closetlike space where she occasionally found stacked crates of onions or jugs of milk left by the kitchen staff when they ran out of space in the freezer next door. Today, there was only the 6-foot-tall metal cabinet containing "the boards," as people called them: the dozens of circuit boards responsible for sending and receiving all the EM 302's pings.

"This thing is very fidgety, like the cables can come loose," said Heffron, shining her flashlight over a series of ethernet cables plugged into the board, "and this is a million-dollar sonar." If something went seriously wrong with the EM 302, this was where Heffron usually went. "Most of the troubleshooting is just gently pulling these things out and sliding them back in and crossing your fingers," she said, demonstrating with the cables. On an expedition last year, one of the boards had failed and Heffron had ended up crammed behind the sonar cabinet for hours. She pointed to a grimy space where she had stood, balancing on a pipe, her head shoved against the ceiling, talking with the other mappers by radio as they went through a series of fixes.*

As if to underscore the antiquated feel, Heffron stepped back, pulled up a steel grate in the floor of the sonar room, and disappeared down a hatch. I passed the flashlight down to her, and she shone the high-wattage beam around a 3-foot-high crawl space. On one side sat the ship's sea chest, a watertight steel box about the size of a microwave oven. Seawater is pumped into the sea chest via an exterior valve, cooling the engine. A sensor inside the sea chest also measured the changing speed of sound in the water drawn in from outside. After each cruise, Heffron would climb down there, open the sea chest, and wipe a thick, viscous membrane of marine growth off the sensor inside. "It's amazing how much stuff grows on it so quickly," she said, holding up the thermometer-like sensor.

My first night on board *Nautilus*, I heard a strange noise just as I was falling asleep. It sounded a little like someone stepping on a rubber ducky. Sound is the enemy of a survey vessel, something I saw, or

---

* E/V *Nautilus* has since upgraded its sonar system by installing a Kongsberg Simrad EC150-3C 150 kilohertz transducer that is the first of its kind to combine an acoustic Doppler current profiler (ADCP) and an EK80 split-beam fisheries sonar into a single instrument.

rather heard, inside *Nautilus*. Any noise on board, from rattling air conditioners to chugging diesel engines, competes with the sonar sounding the seafloor. As we descended inside the ship, the quack I had heard grew louder and louder. It turned out to be the sub-bottom profiler, another of the many sonars on board, which was positioned right next to *Nautilus*'s steel hull. Unlike the multibeam, which sends out a sound frequency that humans can't hear, the sub-bottom profiler emits an audible rising and falling sequence. The multibeam sonar reveals the seafloor beneath the ship, while the sub-bottom profiler goes deeper, revealing what's buried down below in the sediment, whether it's oil reserves, buried cables, or artifacts of a long-lost civilization.

Sound is where an ocean drone has a leg up on any survey ship. Apart from a small diesel engine, the drone glides quietly without even the flapping of a stiff carbon-fiber sail. *Nautilus*'s old, ornery engine, nicknamed Thor for the way it roared to life, had recently been replaced. The new one was supposed to be much quieter—except that after it was installed, the background noise on the ship increased. After much crawling around in the dark confines of the engine room, the ocean mappers identified the noisy culprit: a rattling fuel intake pipe.

The tight spaces, the cramped crawl spaces, the air-conditioned rooms whirring with servers: surveying the seafloor requires so much human labor. An ocean mapper at sea is always being called on to adapt, to problem solve, to jerry-rig a solution, and to wear many hats—literally. One day it's a hard hat, the next a straw sombrero for standing for hours under the hot sun. The first rule of modern seafaring: never wear anything you like, because it will immediately get splattered by oil or paint. (I leaned that firsthand by leaning against a wall that the crew had just painted.)

Erin Heffron chuckled when I told her that ocean mapping seemed surprisingly old-fashioned. That was when she told me about an especially antiquated process she had once witnessed on board a government survey vessel: calibrating a fish-counting single-beam sonar. "The first time I saw it I thought, 'This isn't real. This is really how we do this?'" she said. Initiation rites are common at sea, so there was a real possibility that the crew might haze her. The most famous example is the line-crossing ceremony, during which newbies who are crossing the equator for the first time are hazed into becoming oceangoing shellbacks. The four-hundred-year-old ceremony used to be brutal for new recruits on board military and merchant ships. Now it's mostly a good excuse to put a mop on your head and pretend you're Neptune, the Roman god of the sea. Another hazing ritual more specific to ocean mappers: veterans trick newcomers into donning elaborate safety gear—goggles, helmets, knee pads—before deploying a harmless probe that measures temperature into the ocean. (Heffron was very nice and did not put me through the paces.)

However, as it turned out, the sonar calibration Heffron witnessed was not an elaborate hazing ritual. The all-day process involved a team of scientists weaving a cradle of rainbow-colored fishing line to hold a sphere of tungsten carbide about the size of a golf ball. The sphere was then suspended from multiple outriggers—a heavy-duty winch attached to the gunwale—and dangled dozens of feet beneath the ship's sonar. After much blind maneuvering under the ship, the sonar finally pinged off the tungsten carbide sphere, and the sonar was now correctly calibrated to count fish.

In 2020, NOAA canceled its annual ship survey of the Alaska pollack fishery due to the COVID-19 pandemic and rented three Explorer ocean drones from Saildrone to conduct an acoustic survey instead.[6] The survey, usually performed by a thirty- or forty-person crew on a ship for weeks at a time, was undertaken by one or two operators on land overseeing multiple drones at once. Brian Connon told me that the drone's fish-finding sonar is calibrated in the same arduous fashion and that human labor is still needed for certain aspects of the acoustic survey, like trawling and sampling the fish. But once the ocean drone is up and running, it requires no meals, no breaks, no runs back to land to refuel or restock. The drone's human operators back on land can swap off round-the-clock shifts. That huge drop in the need for human power made me wonder whether ocean mappers were worried about losing their jobs to automation, as so many professions are experiencing today.

Connon fields this question all the time, he said, and the answer is no. "I want to free up [the ocean mapper's] mind to focus on the things that are hard, that require human intervention, and not have to worry about the little fiddly stuff that takes a lot of time," he said. "I think that the job is changing. I got the standard training the IHO has out there. There's nothing in there on using machine learning or artificial intelligence." Very soon, the kind of troubleshooting that Heffron did aboard *Nautilus* could very well become a thing of the past.

The prospect of a flesh-and-blood ocean mapper being replaced by drones and algorithms made me feel a little sad, I'll be honest.* All of the real-world problem-solving on board a ship, the jiggling of wires, the crawling behind circuit boards, actually *being on the ocean* seemed far more interesting than checking on a mapping drone through a phone app. It was the mapmakers behind Seabed 2030 who had first captured my attention. I had once imagined the chart of the seafloor as a sort of tapestry woven by many humans coming together from all points on the globe. Call me a romantic, but that is exactly how maps have always been made. A single map is not a self-contained document but a compilation of "what others have seen or found out or discovered, others often living but more often dead, the things they learned piled up in layer on top of layer so that to study even the simplest-looking image is to peer back through ages of cultural acquisition," writes Denis Wood in *The Power of Maps*.[7] Instead of humans striking out on an expedition, fumbling their way from ignorance to knowledge, robots would do the work for us. Who would willingly give up such an awesome job?

One argument in favor of automation is that it might expand access to ocean mapping. Historically, surveying the seas has been the domain of white men from developed countries. Thanks to the efforts of Marie Tharp and other early female mappers, more women are going to sea today, but they still tend to be white women from the global North. An ocean drone operator, on the other hand, could literally be anyone, anywhere in the world, with an internet connection. Imagine a deaf mapper or a wheelchair-bound mapper getting a chance to contribute. Also, the lower cost of a USV might mean that developing countries

---

* Saildrone clarified that its technology is not intended to replace the ocean mapper but rather to "augment and extend" the ocean mapper's reach. However, an invention can often yield unintended consequences that transform industry and society.

with limited budgets could afford to chart their own shores. So far only industrialized nations can afford to pay for continental shelf surveys and expand their maritime boundaries through the UN's Commission on the Limits of the Continental Shelf. With drones mapping the sea-floor, all that might change.

Larry Mayer, at the University of New Hampshire's CCOM pro-gram, had absolutely zero qualms about mapping the seafloor by drone. "As someone who trains people to map the ocean floor," I asked him, "aren't you worried this technology is going to make mappers obsolete?"

"I would love to make the need for ocean mapping obsolete," he responded, dead serious. Mayer has gone on more than ninety research cruises during his four decades in ocean mapping. Each summer, he makes a trip through the Arctic's Northwest Passage on the US ice-breaker *Healy*. You might think that he would want the adventure to live on, but he seemed to have made enough expeditions to last a lifetime. The romance of spending months at sea, collecting bits and pieces of maps, was all well and good, but now he just wanted the job to be done already and to move on to more pressing problems. "I'm not worried about putting people out of work," he said. "I think we're going to just create more and better maps. That is the starting point, and there's so much else. We're going to understand the Earth a bit better, and we're going to discover mysteries and find ways to solve them."

3.

If Seabed 2030 has a predecessor, it might be the International Map of the World, better known as the Millionth Map. Dreamed up in 1891 by Albrecht Penck, a young geography professor at the University of Vienna, the Millionth Map aimed to chart all the world's landmasses at a scale of 1 to 1 million (1 centimeter to 10 kilometers; 1 inch to about 16 miles).[8] Just as it is astonishing that we know more about the topography of Mars than we know about the ocean floor, it is equally astonishing that in 1891 no complete map of all the world's landmasses existed. By the late nine-teenth century, the tools for surveying land had existed for more than a century. In 1789, France finished the world's first modern topographic survey of an entire country,[9] something that took several decades and generations of the Cassini family of cartographers to pull off.*[10] In 1815, William Smith published the first geological map of Great Britain, meticulously tinting each rock stratum by hand—another world first, and a beautiful one at that. Fifteen years later, the Geographical Society of London (now known as the Royal Geographical Society) was established with the express purpose of charting the world scientifically.

That was at the pinnacle of the imperial age, when colonizing nations were prone to puffing up their real estate on official maps and neglecting to map the rest. That mostly meant the extreme regions on Earth that were too inaccessible, too expensive, or too challenging to survey, and so basic geographic questions remained unanswered: Where was the source of the River Nile?[11] What was the height of Mount Everest? Was Antarctica the world's seventh continent or just a big block of ice?[12] "As late as 1885, it was estimated that not more than 6,000,000 square miles, less than one ninth of the land surface of the globe, had been surveyed or was in the process of being surveyed," wrote the map historian Lloyd Brown. "Many countries, because of apathy and the prohibitive cost of an instrumental survey, had failed to produce an accurate topographic map of their own territory."[13]

In 1891, Albrecht Penck traveled along the Swiss Alps to pitch the Millionth Map at the Fifth International Geographical Congress in Bern. He called it "a common map for a common humanity"—a phrase that echoes Seabed 2030's own quest to see the world's ocean completely mapped more than a century later.[14] Penck's colleagues applauded the idea enthusiastically. Some even wondered why they hadn't thought of the Millionth Map themselves.

The unevenly mapped landmasses were not the only problem that the Millionth Map intended to fix. Another critical issue for cartographers is consistency: using the same symbols, colors, language, units of measurement, and lines of meridian around the globe. Those last

---

* King Louis XIV was disappointed with an early draft that revealed that France was smaller than he had expected. "I paid my academicians well, and they have diminished my kingdom," said the French king with not a small amount of malice.

two—the units of measurement and the lines of meridian—turned into a point of national pride among the imperial powers of the day. The British set their prime meridian at Greenwich and used imperial measurements. The French followed the metric system and used Paris as the prime meridian. Other countries stepped forward to propose their own meridian lines, "most of which were offered without reason except political jealousy, a perverted streak of patriotism or intellectual jaundice," wrote Brown. The only truly neutral territory was the ocean, but of course it would have been too difficult to build and maintain an observatory out on the high seas.[15] All the disagreements over national rules and standards during a time of rising nationalistic fervor foreshadowed the cataclysm of World War I, which would soon engulf Europe.

In 1904, thirteen years after his first appearance in Bern, Penck pushed for the Millionth Map once again.[16] At the Eighth International Geographical Congress, held in Washington, DC,[17] he presented three trial maps he had drafted using his proposed standards. The trial maps convinced the gathered delegates, and the Millionth Map began to make progress, albeit at a plodding pace. The language was set in Latin characters. Greenwich became the international prime meridian—a crushing decision for the French, who were forced to reengrave three thousand maps dating back to the 1720s, when they had founded their first hydrographic office.[18] The British lost the battle of imperial measurements, however, and the scale was set in metric units.

Another decade passed, and in 1913 the delegates gathered once more in Paris. At last, the thirty-five countries represented at the International Map of the World conference reached an agreement on a complete set of rules for the Millionth Map.[19] It had taken Penck more than twenty years to reach a point where the work could actually begin in earnest. A central bureau for collecting the sheets of the Millionth Map was established at the Ordnance Survey's head office in Southampton, England. (Coincidentally, this is only a few miles from the National Oceanography Centre, where Helen Snaith and her team oversee incoming seafloor maps for Seabed 2030.) For a moment it looked as though Penck's dream might become reality at last.

Six months later, a young South Slav nationalist stepped out of a crowd in Sarajevo, raised a pistol, and shot Archduke Franz Ferdinand, a nephew of the Austro-Hungarian emperor, and his wife, Sophie. The assassinations sparked World War I, which would burn through millions of lives in the coming years. The Millionth Map, and its rallying call for cartographic unity, immediately moved to the back burner. Maps were needed to fight wars, not to bring the world together. In 1914, the central bureau in Southampton closed during the war. When it reopened afterward, the Millionth Map struggled to regain its footing.

Throughout the twentieth century, there were fits and starts of enthusiasm for reviving the Millionth Map, followed by long periods of abandonment and gridlock. Albrecht Penck died in Prague at the end of World War II, but his dream to map the world's landmasses lumbered on for another half century. The Millionth Map finally met its end when another map was finished first. After World War II, aerial mapping made huge leaps and bounds, moving from photographing Africa and South America in the late 1940s[20] to sophisticated radar imaging used to survey the entire Amazon basin for the first time in the 1970s.[21] Then the rise of mass air travel created a pressing, profit-driven need to map the world's landmasses as quickly as possible. The aviation industry needed to guarantee that its pilots and passengers wouldn't crash into mountains. In 1989, the Millionth Map came to a quiet end when a UNESCO report recommended that the UN stop monitoring the project as most nations had stopped producing new sheets. It had been supplanted by the World Aeronautical Chart, which became the first continuously updated and complete map of all the world's land.

Perhaps it's not surprising that industry moved faster than scientists and governments. But why exactly did the Millionth Map fail? It was a far more inspiring project than the World Aeronautical Chart—and yet it failed nonetheless. "The progress of the [Millionth Map] is the history of cartography repeated with all its ramifications, a record of man's innate aversion to change, his preoccupation with his own backyard," wrote the map historian Lloyd Brown despairingly in 1949 when the Millionth Map had stalled once again.[22]

Seabed 2030 and the Millionth Map have much in common. Both kicked off at politically fraught moments in history: the Millionth

Map during the rise of nationalism in the early twentieth century and Seabed 2030 during the rise of populism and democratic backsliding in the early twenty-first. Both maps aimed to unite nations behind the universal goal of understanding the planet. Yet the Millionth Map failed not because the task was impossible but because not enough people cared.

4.

If autonomous mapping is the only way to finish Seabed 2030 in time, what did Richard Jenkins at Saildrone think? Could an ocean drone finish the first map of the seafloor that Prince Albert I of Monaco envisioned more than a century ago? Jenkins had done a few back-of-the-envelope calculations on that exact question. "With twenty Surveyors, you could map the world in 9.6 years," he told me. At that moment, in late 2021, there was just barely enough time to meet the deadline, if you ignored the fact that twenty Surveyors did not yet exist. Only one did. Hypothetically, Seabed 2030 was still possible. Realistically, it seemed doubtful.

Jenkins had done a few other rough calculations on whether Seabed 2030 would be possible *without* ocean drones. "It's absolutely not possible. It's just not. It's too expensive. It will take far too long. We don't have enough ships on the planet to do it. So I call bullshit," he said bluntly. Then he backpedaled a bit: "I admire the motivation and the energy [of Seabed 2030]. . . . We've spent tens of millions investing in the solution to get it done. It's not like I'm not a believer, but unless it's a Saildrone-type [vehicle] that does [Seabed 2030], it's not happening. Just drop the illusion and call it Seabed 2090."

The year 2030 was always an arbitrary deadline for finishing the first map of the world's seafloor. There is no fee or penalty if we blow the deadline, except perhaps all the unknown discoveries we'll miss out on. It took nearly a century to survey all the world's land, and the mind-blowing revelations unearthed along the way were more than worth the effort. We've since settled long-standing debates in geography: Antarctica is indeed a continent, not an island or an ice sheet. The source of the River Nile is largely believed to be Lake Victoria, although some competing tributaries are still tossed around.

And then there are the discoveries that we never even dreamed we would make. Satellites revealed an uncharted 125-mile-long island in the Canadian Arctic in 1976.[23] It was dubbed Landsat Island, after the satellite that had mapped it. A year later, an airborne survey of the Guatemalan rain forest revealed a network of canals, the first solid proof that Mayans had developed sophisticated agricultural systems long before Europeans arrived.[24] Seabed 2030 is far more ambitious than the Millionth Map in sheer territory alone. There's no telling what we might uncover while mapping the sunken half of the planet's surface. So what will we be missing if we fail to finish Seabed 2030? I flew to Florida to find out.

CHAPTER 9

BURIED HISTORY

1.

Over the years I spent writing this book, I kept a list of seafloor discoveries and phenomena. Pretty soon, the list grew so long that I had to condense it to just a few of my favorite highlights:

In 2017, an international research team confirmed the world's deepest-dwelling fish, the Mariana snailfish (Pseudoliparis swirei), *living up to 26,246 feet deep in the Mariana Trench.*[1]

In 2018, E/V Nautilus discovered an "octopus garden" with a thousand female octopuses brooding oblong white eggs in an underwater hot spring 2 miles deep off the coast of California.[2]

In 2020, the Australian marine geologist Robin Beaman and the rest of the team aboard Schmidt Ocean Institute's R/V Falkor discovered a coral reef taller than the Empire State Building in the Great Barrier Reef.

In 2021, a team of scientists landed on an uncharted island in the Arctic Ocean that turned out to be the world's northernmost landmass.[3]

In 2021, scientists aboard the Alfred Wegener Institute's icebreaker R/V Polarstern discovered a gigantic icefish colony a thousand feet deep on the Antarctic seafloor, with some 60 million active nests spread over 92 square miles.

In 2022, an international team discovered the final resting place of the legendary explorer Ernest Shackleton's ship Endurance *off Antarctica.*[4]

In 2022, UNESCO and an ocean-mapping group found a pristine deepwater coral reef off Tahiti.[*5]

The most fascinating thing to me about this list is that most of these findings were happenstance or accidental. The team that found the world's northernmost island, for instance, realized it only months later when they looked closer at the map. No one could say precisely why the island hadn't been charted before. It might have been covered by pack ice or thrust up above the waterline by storms and sea ice razing the seafloor, the resulting feature known as an ice pressure ridge.

Wild and random things such as this happen all the time in ocean science. It's one of the few fields in which major groundbreaking discoveries are not only possible but probable. "The next generation, the generation of explorers that are in middle school right now, [they] will explore more of Earth than all previous generations combined," Bob Ballard has said, "and they're the ones that are going to unlock these two million, three million years of human history that's preserved in the deep sea."

The United Nations estimates conservatively that 3 million shipwrecks are out there, undiscovered, on the seafloor.[6] Every year more ships go down, and less than 1 percent of those wrecks have been explored. During the search for the missing Malaysia Airlines flight 370,

---

* A controversy later erupted over this discovery or, rather, *who* gets to "discover." Local fishers and divers in French Polynesia pointed out that they had known about this deepwater reef for quite some time; the researchers involved said that their "scientific discovery" was mischaracterized in the press. Thomas Burelli, an environmental law researcher at the University of Ottawa, told *Hakai* magazine that the incident evoked terra nullius attitudes: "not mapped, doesn't count." See https://hakaimagazine.com /news/discovering-what-is-already-known/.

which mapped more than 279,000 square kilometers (107,722 square miles) of the Indian Ocean, two nineteenth-century shipwrecks turned up—again, a completely accidental discovery.[7]

Shipwrecks are just one part of the human story that we might uncover when Seabed 2030 finishes a complete map of the seafloor. For 90 percent of human history, there was much more land on Earth than there is today. The glaciers of the last ice age reached their peak around twenty thousand years ago (a period known as the Last Glacial Maximum) and then began to melt, releasing freshwater back into the ocean, raising sea levels, and drowning continental shelves. From 5.8 million to 7.7 million square miles of coastline slipped below the waves—a lost Atlantis that is roughly equal to the size of South America.[8] Depending on geography, some places lost more land than others. The ocean swallowed 40 percent of Europe's landmass, and Denmark is now one of the leaders in exploring the underwater world of submerged archaeology.[9] The state we now know as Florida also used to be far larger than it is today, having lost roughly half its land since the last ice age.

Florida has a long history of underwater peat bog burial sites being discovered on land, as in 1982, when construction workers dredging a pond near Cape Canaveral turned up human remains that led to the discovery of an Archaic Period cemetery. More than half of the 168 bodies there were so well preserved in the oxygen-free peat that they still had brain matter in their skulls.[10] Then, in 2016, a diver near Venice Beach, Florida, pulled a human jawbone from the seabed. That led to the discovery of a 7,200-year-old Early Archaic Period cemetery about 300 yards offshore.[11] The Manasota Key Offshore site became the first pre–European contact burial site discovered in a purely marine environment in all of the Americas and one of just a handful known around the world. The discovery put Florida's offshore archaeology onto the map.

Archaeologists have historically focused on land sites that are easier to access and explore. "One big misconception that [traditional] archaeologists have is that when the sea level rose, it just erased everything," said Shawn Joy, a Florida archaeologist who specializes in sea level rise and underwater precontact sites off the Gulf of Mexico. Joy is part of a new generation of scuba-diving archaeologists who are using the tools of ocean mappers to track down underwater sites on the seabed.

When I met Joy and his collaborator Morgan Smith, an assistant professor at the University of Tennessee at Chattanooga, in the summer of 2021, they had just finishing mapping portions of Florida's Apalachee Bay in the Gulf of Mexico, where they turned up nearly two dozen potential sites from the Late Archaic Period (5,000 to 2,500 years ago).

Finding shipwrecks is pretty straightforward, they say. Locating a precontact site buried on Florida's sunken continental shelf? Now, that's as tough as it gets. The Manasota Key Offshore discovery brought new attention to their quest, but Smith and Joy still find it hard to communicate just how much history is still out there buried on the continental shelf. Around 40 percent of the world's population today lives within 100 kilometers (60 miles) of a coastline.[12] In Florida, that statistic is nearly double, with 76 percent of its population, or around 15 million people, living by the coast. Throughout antiquity, people have always been drawn to water, whether a river, a sinkhole, or a coastline. "Put people on the coast, and watch the life just rev up: hunting, gathering, art, culture . . . ," said Joy. "Get people inland, and it's like 'Well, shit, how am I going to live today?'" He gave a rough Neanderthal shrug. If even a fraction of early Floridians liked the ocean as much as present-day ones do, there's a whole world of human history waiting there to be uncovered. And the farther offshore you go, the further back in time you travel.

We know very little about the early inhabitants of Florida, but we do know that as the glaciers of the last ice age melted, they pumped water into the Gulf of Mexico and raised its water level, forcing early Floridians to retreat inland. Thousands of years later, the seas are rising once again, this time as a result of humans burning fossil fuels, threatening the homes and livelihoods of more than 400 million people worldwide.[13] That makes investigating the drowned history of Florida, and other sunken landscapes around the world, all the more pressing. What happened to the people who lived here on what is now seabed? What can they teach us about navigating a sinking world? "Even people

who believe in climate change and understand sea level rise, it's hard
to wrap your head around," Morgan Smith told me during my visit to
Florida. "All this was dry land," he said, gesturing across the Apalachee
Bay. "Imagine if Florida just continued."

2.

X marked the spot. We were close.

That morning, a crew of diving archaeologists boarded a boat in
the town of St. Marks, Florida, and steamed out to the Gulf of Mexico.
As the sun rose and the morning fog lifted, the pontoon boat followed
the winding St. Marks River through the mangroves. Alligators' eyes
sank into the brackish water as the boat chugged past, noisy herons
took flight. On our right we passed Fort San Marcos de Apalache, built
by the Spanish in the seventeenth century at the confluence of the St.
Marks and Wakulla rivers. Over the centuries, the fort was burned,
rebuilt, looted by pirates, occupied by the British, retaken by the Spanish, and later seized by Andrew Jackson, the seventh president of the
United States.[14] What had once been a commanding fortress, hurricanes, and history had eroded into an overgrown spit of swamp lined
by boulders. Here and there, tall copses of pines and pond cypresses
rose above the swamp; each one was likely an archaeological site with
an oyster midden buried at its base. This is the "real" Florida, as locals
here in the Big Bend like to say, far from the glitz and glamour of Mi-
ami but rich in history.

Up ahead, a postcard-perfect white lighthouse announced our entry into the Apalachee Bay. It was a calm day on the gulf, not a ripple
brushing the flat, limpid surface. Two weeks earlier, Joy and Smith had
spent all night here on Joy's sailboat, running a sub-bottom profiler—a
sonar similar to the one I had heard quacking against the steel hull of
the E/V *Nautilus*—back and forth over the bottom. They had "mowed
the lawn" that night, going back and forth, back and forth in search of
underwater sites. The sites they had found, eighteen in all, were sprin-
kled across Joy's GPS screen as he steered the boat. Over the coming
days, we were hoping to dive them all.

"Twenty-five meters," Joy called from the helm. We were 2 miles offshore now and closing in on the first site. Joy killed the engine. For a few silent moments, the boat coasted on momentum. I leaned out over the gunwale, looking down through shallow green water. Rays of sunlight lit up the swirling underwater pastures of bright green eelgrass, big stumps of coral, and rippled sand dunes dotted with pretty white scallop shells.

"Ten meters," Joy called, watching the GPS from behind neon-green shades.

"What side?" Smith called back excitedly. He was standing on the bow, ready to fling the anchor he held in his arms as soon as Joy gave the word.

"Port," called Joy. Smith moved to the left side of the boat. All of us on board—Smith, Joy, a volunteer diver named Ximena, and I—peered into the water, searching for a sign of the site. Sand, eelgrass, coral: it looked the same as everything else we had passed.

An odd shape in the soundings of a sub-bottom profiler had brought us to the spot. A sub-bottom profiler typically scans the sediment of the seabed, hence the "sub-bottom" in the name. The results look a little like a drippy layer cake in gray scale, each layer representing a different slice of sediment. Oil and gas surveyors use sub-bottom profiling to find untapped reserves; ocean mappers use it to spot buried cables or locate sturdy terrain on which to build bridges. Why not use it to try to find ancient stone artifacts, too? Well, archaeologists had tried that before, with little success. In 1982, the world-famous Danish bassist Hugo Rasmussen took sixteen Stone Age flakes and blades to the Bang & Olufsen sound laboratory. In a series of experiments, he found that each artifact produced a specific tone when blasted with sound frequencies. Experts reasoned that it should be possible to track down underwater sites using sound, except they couldn't quite figure out how to do it.[15]

In 2014, a team of archaeologists and geophysicists from Israel and Scandinavia joined forces to see whether sub-bottom profiling might be the missing key. Running field experiments at two confirmed underwater sites in Israel and Denmark, they discovered odd shapes in the sub-bottom soundings. The shapes were hovering in the water column, fat at the top, tapered at the bottom, and looked a little like haystacks, so that was what they called them. The haystacks didn't move, so they weren't a school of fish, and they were found floating right above stone artifacts buried in the seafloor. It turned out that the sub-bottom profiler *could* reveal underwater archaeological sites, but the experts were looking for signs in the seabed where they expected to find the artifacts rather than in the water above.[16]

Hollywood movies tend to portray Stone Age people as grunting, feral cave dwellers, but the tools they made tell a different story. They had a sophisticated appreciation of stone and sound, carefully selecting the right material for arrowheads or knives by tapping the stone first to see whether it made the correct sound. The right tone told the early toolmaker that the stone was free of flaws and fissures: a quality piece that would hold up long after he moved on from the quarry. The aural quality of that stone is a clue left behind for ocean-mapping archaeologists thousands of years later. Here in Florida, Smith and Joy were using the technique, now known as the Human-Altered Lithic Detection (HALD) method, to lead them directly to ancient sites sunken on the seabed across the northeastern Gulf of Mexico.[17]

"Huck it! Huck it!" yelled Joy. Smith sent the anchor flying as far from the boat as possible. A splash. It sank. The boat ran to the end of the line and then slowed to a gentle stop. Smith, still wearing his khaki shirt and pants, dived in immediately. The sound of water breaking snapped me out of my early-morning daze. Right, of course, this wasn't a fun boat trip. We were here to dive.

Smith paddled around the boat for a moment, head down, goggles on, scanning the bottom. "Oh, yeah," he said, raising his head out of the water, sopping bangs plastered to his forehead. "Artifacts everywhere." He did a jackknife down to the bottom: flippers high in the air followed by a slow slide to the bottom. A moment later he was back, holding something small in his hands. He pulled his goggles off and studied the object a little more closely.

"What is it?" I asked eagerly, leaning out over the side of the boat.

It was chert, a prized material among the early stone-working people of the world. By whittling away at a core of chert, the early inhabitants of Florida made a range of notched and pointed blades, scrapers, and choppers in a process known as flint knapping.* The flake Smith was holding was the leftover debris. He swam over to the boat and dropped the flake onto my outstretched palm: the first find of the day. The jet-black stone was about the size and shape of a tortilla chip. One edge was about an inch thick; the piece sloped down to a wafer-thin triangular tip that was almost transparent when I held it up to the dazzling Florida sun.

Apart from Smith, the last time someone had touched that flake had been more than three thousand years ago. The thought made me shiver. There's something addictive about being "the first": to see something, touch something, do something that no one else has ever done. It was the same feeling Victor Vescovo had felt while exploring the Puerto Rico Trench for the first time. The flat seascape at the deepest point of the Atlantic wasn't what had thrilled Victor the most; it was the knowledge that he had arrived *first*. The same urge drives so much exploration, so much filling in of the map. There's also something possessive about claiming to be "first." It reminded me of the way phrases such as "virgin territory" make me squirm.

Of course, it wasn't as though I were Howard Carter, peering into King Tut's tomb. I was holding a rock that looked, to the untrained eye, like any other rock. How would I know that this rock had been left by a precontact hunter hacking away at a boulder three thousand years ago? "Hertzian mechanics," said Smith. He was hanging off the ladder on the boat's stern. When a stone is struck, it behaves a certain way, he explained, "like a bullet shattering a pane of glass." He took the flake from me, ran a finger along the sharp straight side. "This is called a flat form, where the object was struck. And you feel here, right beneath that platform," he said, pointing to the chunkier part of the stone, "you can feel how the stone expanded there, where it's bulbous. The only way that happens is by something very hard striking the stone."

---

* On YouTube, there is a surprising number of channels devoted to flint knapping and other so-called primitive survival skills that include start-to-finish instructions for how to knap a complete stone tool kit.

Both Smith and Joy are trained in geoarchaeology, using geological principles and methods to confirm how a site formed, when it might have been abandoned, and why an artifact was deposited there. This deeper understanding of oceanography and marine geology is critical for convincing the skeptics of submerged archaeology. Some coastlines do stir up enough energy to create "geofacts," flakes that break off naturally.[18] In California, sudden downpours cause parched rivers to swell and burst with enough energy to create geofacts. Archaeologists at the Calico Early Man Site in the Mojave Desert claimed to have found stone artifacts dating back fifty thousand years or even earlier, but those findings were later disputed because the rocks had been found in the geological deposits of a fast-moving river.[19] But that is not the case in the Gulf of Mexico, explained Smith. It's far too sluggish to produce geofacts.

What about hurricanes? I asked. Couldn't a category 5 generate enough power to smash stones apart? "When hurricanes hit, oh, it gets *nuts* out here," Joy piped up from the helm. During 2018's Hurricane Michael, the last "big one" in Florida's Big Bend region, winds and waves had surged into St. Marks, ripping out trees and roads, downing power lines, flooding homes. Surprisingly, the bottom of the Apalachee Bay hadn't shifted all that much. Joy told me about a site he had dived where he had found two stone flakes sitting next to each other that fit together, what is called "refit," meaning the two pieces had been hacked off each other. Joy had set the two pieces down on the seabed next to each other, right where he had found them. About a year later, he had returned and found them lying in exactly the same position. "We're trying to disprove that misconception [against submerged archaeology] by saying 'Hey, look, we have integrity here,'" said Joy. "This stuff hasn't moved all that much since sea levels rose."

In short: human hands made the stone I was holding. The Apalachee Bay appears to be a drowned quarry site where precontact people worked and lived. "When you find the quarry, now you have an anchor on this landscape. Because rock is really heavy, right? You don't want to carry it far. So somewhere around here, there should be a base camp," said Smith. Finding a base camp could lead to discovering a collection of stone tools that could lead to a better understanding of how an ancient people lived. But the starting place was the flake I held in my hands.

3.

How did people populate the Americas? That's a contentious question in archaeological circles today, but many underwater specialists believe the answer is written on landscapes swallowed by the sea. For decades, the story most people learned in school—the Clovis-first theory—held that sometime around 13,000 years ago, a single group of people from Asia migrated into the Americas along the Bering Land Bridge. As the seas lowered during the last ice age, a land crossing rose out of the ocean and briefly connected Siberia and Alaska. Once those first people made it to the Americas, they worked their way south along an ice-free corridor that opened through the interior of North America.[20] The Clovis-first model aligned with the ancient sites that were first discovered inland, such as the famous 13,000-year-old site near Clovis, New Mexico, that gave the early people their name.[21]

Throughout much of the latter half of the twentieth century, the Clovis-first theory reigned supreme. Just as continental drift was fiercely rejected until Marie Tharp's maps showed otherwise, the Clovis-first theory was fiercely defended. Up-and-coming archaeologists were discouraged from pursuing alternative ideas lest they stunt their budding careers. Sites that didn't match the model were disputed; sites that did were accepted.[22] Just a half hour's drive from the Apalachee Bay is one of those long-disputed sites, buried in a sinkhole along the Aucilla River.

As early as the 1960s, hobby divers started exploring the Aucilla. Its waters as muddy as chocolate milk, they swam with alligators and manatees and pulled out mastodon teeth as big as car tires, along with the bones of ancient bison and camels. Locals knew that there was history buried there, but they didn't know quite how old until the discovery of a 14,550-year-old site preserved inside a 30-foot-deep sinkhole known as Page-Ladson.

Florida sits atop a flat limestone platform that is honeycombed with sinkholes like Page-Ladson. Over time, the holes grow bigger and weaker due to the effects of the trace amounts of acid contained in rainwater. Eventually the earth gives way to the water, swallowing a house, an apartment building, or whatever else happens to be standing there at the time. In modern Florida, the sinkholes are a major home insurance issue, and they have killed people, too. In early Florida, however, sinkholes were lifelines: they provided a watery oasis on the dry, dusty savanna and attracted megafauna and people. The sediment in the sinkholes also preserves a snapshot of the place. In 1983, a team converged at Page-Ladson, and over the next fifteen years rotating volunteers, hobby divers, students, and professors systematically excavated the sinkhole and uncovered the site's crowning find: a mastodon tusk with long grooves scratched along its side that looked like the work of human hands.[23] But due to the dominance of the Clovis-first theory and the site's position in the Southeast, where no one had expected to find humans who had lived that early, Page-Ladson's provenance was doubted for years.

In 1997, the discovery of a site in Chile struck the definitive blow against the Clovis-first theory.[24] Preserved in a peat bog at the sub-Antarctic tip of Chile, the submerged Monte Verde site holds a truly astonishing collection of artifacts: remnants of wooden structures and animal-hide walls, remains of meals of wild potatoes and berries.[25] Carbon-dated to 14,800 years old, Monte Verde predates the earliest Clovis sites by more than a millennium. That led to a wholesale re-evaluation of long-dismissed pre-Clovis sites such as Page-Ladson,[26] which is now confirmed as being the oldest radiocarbon-dated site in the Southeast, at 14,550 years old.[27] Most important, Monte Verde created an exciting paradigm shift in which the Clovis-first theory no longer dominates the debate. On this playing field, the traditional theory of the Bering Land Bridge migration dukes it out with three possible alternatives. Each of the alternatives features a transoceanic or coastal migration in which submerged archaeology could play a major role in settling the debate.

The Kelp Coastal Highway theory holds that a Pacific coastal migration took place along the Bering Strait between northeast Asia and northwest America. Moving by foot or boat, early humans foraged off familiar marine life found in the kelp forest. When the sea level rose after the last ice age, the ocean covered their tracks. The Transpacific Crossing theory proposes that the voyage happened by boat, perhaps between Asia and South America, and again, sea level rise might have buried the evidence.[28] The third alternative, the Solutrean hypothesis, suggests an eastern entry, with hunter-gatherers from Europe navigating along the ice pack into North America between 18,500 and 20,000 years ago.[29] However, many experts regard the Solutrean hypothesis as impossible, based on archaeological and genetic evidence, as well as dangerous, given its popularity among white supremacists.

"If we're going to actually get an answer on these [alternatives], we're going to have to look underwater," said Morgan Smith as we rode around the Apalachee Bay, checking out more sites. "Whatever it is, it's gotta be underwater, and we're not going to know what really happened until we survey the area." However, the seabed we were exploring that day was not about answering one of the biggest questions in American archaeology. Smith and Joy were diving to answer a deeper question, one often lost in the vehement debate about "who" populated the Americas first. Namely: What was life like in early Florida before sea level rise and European colonization?

In the official accounts of Florida history, there is often a single paragraph, maybe two, devoted to the precontact period, the thousands of years of human history before Europeans arrived. The Apalachee, after whom the bay is named, were once a major tribe in the region, and their history there dates back at least a thousand years. They had a reputation as fierce fighters and prosperous farmers who played a type of ball game that was both a religious ceremony and a sport. They ate copious amounts of shellfish, and inland Tallahassee residents often unearth the shells of ancient oysters and whelks in their backyards.[30]

When the Spaniards arrived in 1528, they made contact with the Apalachee living in dispersed farming villages and numbering between fifty thousand and sixty thousand.[31] The Europeans were unimpressed with the way of life they found and with the stone tools that Europeans had long before discarded during the Iron and Bronze Ages. "Since there was no way for the explorers to relate the significance of these 'esoteric' practices to their own worlds, their written records include only condescending and uninquiring reports of the aborigines' manufactures," writes the Florida archaeologist Barbara A. Purdy.[32]

In Florida today, the Apalachee have all but disappeared, expelled in successive waves of violent eviction, enslavement, border skirmishes, and the introduction of deadly foreign diseases. A map of southeastern America from the 1720s shows the Apalachee living in a contested borderland between the French to the west, the Spanish to the south, and the British to the north.[33] In 1704, a British raid on a Spanish mission killed and captured thousands of Apalachee, with a few Tocobaga members among them. All were shipped to South Carolina as slaves.[34] A few Apalachee escaped and fled west to French-held Mobile, Alabama. Another group held out a little longer at a Spanish mission near Pensacola. They, too, were forced out in the 1760s and evacuated to a town north of Veracruz, Mexico.[35] Any remaining Apalachee in Florida would have been swept up by the Indian Removal Act, signed by President Andrew Jackson in 1830. The act enabled the federal government to seize land from Native American tribes in the Southeast and force them to march west on a genocidal migration now known as the Trail of Tears. Today, a small band of surviving Apalachee lives two states over in Louisiana, where they've been petitioning for federal recognition since 1996.[36]

As the Apalachee disappeared in Florida, so too did their stone-working skills, uniquely adapted to the chert found in the region. There is a heartbreaking account (one of countless many) of a captured member of the neighboring Timucua tribe attempting to chisel off his iron chains with a stone tool.[37] The stone quarries scattered across the bottom of the Apalachee Bay reveal a glimpse of the stone-working tradition right before the dawn of the Apalachee period. They also capture the skill level of the toolmakers, how a group survived and used the environment, how they made decisions about the material they used, and how they organized labor and laid out work sites with distinct stations for separate activities. Flakes with tiny nibbling marks along the sharp edge reveal that someone used a flake once or twice to slice something and then tossed it away.

Morgan Smith knows firsthand just how sharp those flakes can be. A few years ago, during an archaeological exercise not far from

Apalachee Bay, he was sledgehammering apart a boulder when a flake flew off and sliced his hand open. His supervisor whisked him to the emergency room. Smith still has a piece of flint lodged in his hand, along with some scars from stitches.

As illuminating as the quarries are, Smith and Joy have to be careful about inferring too much from one type of site. Archaeologists are sometimes criticized for drawing too many cultural conclusions from stone artifacts. Partly this is a function of durability: stone is studier and more likely to last in the archaeological record, especially in an offshore environment, than softer materials are.[38] So when I tried to encourage Smith to paint me a portrait of what life might have looked like three thousand years ago, he was cautious, circumspect. "We only have those quarry sites [on the continental shelf]," he said. "We're not seeing what they ate, how they organized, other things they used, like bone and wood; it's just the stone." But a broader selection of sites and artifacts recovered from the seafloor might reveal what life was like in early Florida, before the onset of European colonization and rising seas.

4.

With an oxygen tank strapped to my back and a breathing regulator jammed into my mouth, I dropped into the water feeling as heavy as a stone. The ocean caught me, that bathtub-warm water of the Gulf of Mexico, not refreshing but very, very comfortable. I released the air in my vest and began to sink, salt water closing over my head. The bright sun cut through the surface, sending a prism of light scattering over my arms and legs. The weightless floating, the sunlight filtering from above—it felt heavenly down there. How could anyone ever imagine the seafloor as a dark, hellish place?

Fifteen feet below, Shawn Joy began the day's work by setting a datum on the seafloor: a ground zero around which the site would be mapped out on an X-Y grid. This site's datum was a pin set next to a clump of yellow and electric-red coral growing from a rocky outcropping. He hovered close to the coral and waved an open palm around the base of the rock, and a puff of sediment flew up like smoke. This is hand fanning, the underwater equivalent of the soft brush archaeologists use on land to dust off fragile artifacts. When the dust settled, a small cluster of black chert flakes appeared like magic, right at the base of the coral. Joy glanced at his dive-computer watch and scrawled the coordinates on an underwater notepad strapped to his wrist. Then he slipped the artifacts into a mesh bag hanging from his waist.

Archaeologists sometimes compare documenting a site to investigating a crime scene. In the same way that you wouldn't casually pick up a bullet casing or wipe away blood splatter before the police arrive, the less disturbed an archaeological site is, the better. An archaeologist needs to collect a holy trinity of data to confirm that the site has seen genuine human activity: context, artifacts, and secure dating. With a growing collection of artifacts in his mesh bag, Joy had already met one requirement. Next up: secure dating.

There things got tricky. Chert doesn't contain carbon, so it can't be radiocarbon-dated. Shawn Joy wrote his master's thesis at Florida State University on improving sea level–rise modeling in the Gulf of Mexico by radiocarbon-dating the surrounding coral and applying new statistical techniques. He estimated that the area was last used by people no earlier than three thousand years ago.[39]

Joy unclipped a large spool of yellow measuring tape from his belt and gave the end tab to Ximena, the volunteer diver. She held it close to the coral outcrop, then he swam off away from Ximena in the opposite direction, the measuring tape unfurling behind him. I followed him closely, not wanting to miss a minute and also nervous about being left alone down there. The archaeologists mentioned that playful pods of dolphins sometimes swam around them as they worked. That sounded nice. I hoped that would happen, because I also knew that all three of the world's most dangerous sharks—the tiger, the bull, and the great white—pass through these waters and that Florida has more shark attacks than anywhere else in the country. The heavenly feeling I'd had when I'd first dived into the water quickly faded. Now I was glancing around anxiously, peering into the murky, impenetrable gloom. Just outside the 20-foot circle of visibility, the view closed off. I pushed away the persistent thought of a shark emerging from the darkness—why is the shark always grinning like an evil clown in this nightmare?—and kicked harder to catch up with Joy.

Once Joy reached the end of the measuring tape, he retraced his path, winding the tape back in again. He slipped any artifacts he found along the way into the mesh bag, their coordinates noted. At certain points, he thrust a chaining pen into the seabed to measure just how much sediment lay over top the bedrock. That could help with the third and final requirement: context. On land, archaeologists confirm the environmental surroundings of a site by stripping away sediment layer by layer. The process underwater is similar, but when artifacts lie exposed on the seafloor like this, there's not enough context to date them and thus they need deeper documentation of the underlying seabed.

Up until the turn of the last century, submerged precontact sites in North America were often dismissed by terrestrial and shipwreck archaeologists as too expensive, too technical, and, most important, not worth documenting because sea level rise had probably destroyed them. "If you said 'underwater archaeology,' everybody thought shipwrecks. It's still that way today," said Michael Faught, an archaeologist, former researcher at Florida State University, and mentor to Joy and Smith. In 1986, Faught, then a graduate student, went on his first dive in the Apalachee Bay. Back then, students were towed facedown behind a boat to scout out promising sites on the seafloor. The boat's operator couldn't drive too fast or the student's goggles would fly off, but it was way faster than swimming. "We didn't know what we were looking for," he said. "When I was being towed, it was like grass and grass and grassland, deeper, a sandy bed, deeper grass and grass and grassland shallow, and I was like, 'Oh, crap, that's it. That's a paleochannel.'"

A paleochannel is an ancient river that drowned as the glaciers melted and seas rose. Archaeologically, paleochannels are important because ancient people gravitated to rivers, walking along or between waterways. Before the HALD method was developed, archaeologists found submerged sites by following a paleochannel offshore and looking for sites nearby, a technique known as the "Danish method" after the researchers who developed it. The Danish method was better than blindly towing an archaeological student back and forth across a bay, but it still left a lot of seafloor to cover. "You're really just guessing at the end of the day," said Smith. "Just because there's a river there doesn't mean people were there."

Back in the 1980s, when Faught first set out in the Apalachee Bay, he had trouble keeping track of his sites. GPS was not yet widely available, so it was difficult to record spots once they were discovered. "The first year we went out [in the Apalachee Bay] in 1986, I lost sites," he said. He tried to use dead reckoning: triangulating his position from landmarks on shore. But, he sighed, "that stuff just didn't work." The researchers left behind stakes to mark their work, but they lost those, too, sometimes in the middle of working on them. Faught remembered dropping lighted pontoons onto a site to mark it. The next morning he returned. "We dive in," he said, "and it was like, 'What happened to the site?'" The pontoons had drifted in the night.

At the rest of the spots we dived that day, a pattern quickly established itself: Joy counted down the meters to the point on the map, Smith threw the anchor, and at every site, the anchor landed right by an outcropping mined by precontact people. During one dive, Joy followed the anchor line down to the seafloor and found the anchor snagged right on top of a pile of flaked chert. "That was easy," he scrawled on his underwater notepad.

The HALD method makes it look easy; however, Faught's stories show that it took decades of refinements to get to that point. In the late 1990s and early 2000s, he also used sub-bottom profiling to identify quarry sites and paleochannels, but it didn't have the same precision as the HALD method does today. Archaeologists aren't dragged around behind boats anymore, either. ("Well, that turned out to be dangerous, and you don't get to do that kind of thing [anymore]," said Faught with a chuckle.) But Joy and Smith did have an underwater scooter on board for scouting out sites—"Basically the most fun you can have underwater," as Smith put it. Shaped like a small torpedo with a fan on the front and handlebars on each side, the scooter looked like something straight out of a Jacques Cousteau film. I had to try the thing out. When I dropped into the gulf on the scooter, revving the handlebars furiously, bracing myself for liftoff, I sank to the bottom like a stone; the battery wasn't charged.

Shipwrecks might always hog the spotlight in Florida archaeology.

It is hard to compete with the Treasure Coast on the Atlantic, where a fleet of Spanish treasure galleons went down off Vero Beach during a hurricane in 1715. After the storm, gold and silver coins washed up on shore, and to this day treasure hunters still scour the sand for lost doubloons. However, the Manasota Key site found off Venice Beach did help dislodge some prejudices against submerged archaeology. "That was really, really good for submerged prehistory, because that woke everybody up to it, myself included," said Michael Faught. He's envious of Smith and Joy, who have better tools and techniques for exploring the seabed than he did. "Those guys get to keep building on the known to the unknown. They're going to find some cool stuff." He's hopeful that his successors will use the new tools to push farther offshore and explore further back in time, perhaps even to the Paleo-Indians, the first people to inhabit the region more than ten thousand years ago.

At conferences, Smith and Joy still field questions from skeptics. How can they possibly confirm archaeological sites that have been sitting on the seafloor for thousands of years? But they have also noticed that their talks are drawing in bigger, more inquisitive crowds. Barbara Clark, a regional director of the Florida Public Archaeology Network (FPAN), sees that enthusiasm growing, too. "When I tell people about [Manasota Key], you can see their eyes widen," she said. However, more interest and attention can be a double-edged sword. On our way into and out of the Apalachee Bay, we passed fishing boat after fishing boat, each equipped with fish finders that could mark the same rocky outcrops we were diving. At one site there was an indentation in the sand left by a diver's fin. At another, the broken-off gunwale of a fishing boat rested on the bottom. When the tide in the bay goes out, locals like to party on the sandbars, strolling around the seabed, red plastic cups in hand and salt water up to their ankles. They might not be hunting for artifacts, but what's to stop someone from snagging one?

5.

"It's not lost on us that none of these sites have actual tools at them," said Shawn Joy rather cryptically on our last day diving the Apalachee

Bay. The sun was going down, and the diving team had gathered for a farewell dinner at the Riverside Cafe in St. Marks. All of us were sunburnt, drinking light beers, toasting to three days of successful dives. No one could say that the summer's fieldwork hadn't been a success. The method had worked perfectly, confirming nearly every new site in the Apalachee Bay. Joy and Smith had collected bags upon bags of artifacts, all annotated and ready for analysis. The team had never expected to find a diagnostic: an artifact with irrefutable evidence of human use. Quarry sites were used mainly for cracking rocks apart, so it's unlikely that someone would have left a finished tool mixed in with mining debris. However, the lack of diagnostic artifacts *was* curious, Joy said. It might signal that someone else had gotten to the sites first and snatched up the better artifacts before them.

Florida has a long, complicated history of looting. Just a few months after the Spanish galleons went down off Vero Beach in 1715, looters, pirates, and enslaved divers descended, absconding with a wealth of coins, cannons, and guns. Here in the Big Bend, there's another brand of treasure hunter at work. Tips from the public trickled in: reports of trucks parked outside state parks at odd hours of the night, of men emerging from the brush at dawn carrying shovels and bags. Then one morning in 2013, agents with the Florida Fish and Wildlife Conservation Commission struck, raiding six homes and rounding up thirteen people across north Florida and Georgia in a sting dubbed "Operation Timucua" after the local Indigenous People. The group turned out to be selling Archaic Period and Paleolithic projectile points online and at trade shows. Some were selling for as much as $100,000,[40] others for as little as $15. Leading up to the raid, undercover agents had tracked the members closely. The agents had befriended them, bought from them, and secretly recorded them before arresting them.[41]

The same day as the raid, state officials held a press conference on Operation Timucua at the Museum of Florida History in downtown Tallahassee. The reaction was unimpressed, even hostile. "A defendant allegedly sold a box of about 90 assorted artifacts to an undercover agent for a grand total of $100—not quite Pablo Escobar territory," wrote an opinion editor in the *Tampa Bay Times* after Operation Timucua. ". . . Might this dragnet have been better dubbed Operation Overblown?"[42]

Both the public and the media seemed more concerned about the taxpayer dollars spent on cracking down on a few guys collecting rocks. Even though the heavy-handed tactics of Operation Timucua are fairly typical of US law enforcement, those tactics are not typically leveled against white men without criminal records pursuing what many considered to be a harmless hobby. Operation Timucua was also controversial among collectors because Florida used to run a program that allowed amateurs to keep artifacts found on public land as long as they reported the find and a state archaeologist approved it. But the program shut down in 2005 after it became clear that too many people never reported their finds.[43]

Barbara Clark had zero sympathy for the accused. "The people who were arrested for Operation Timucua were making six figures," she told me. "They had big boats, fancy equipment, things like that. They weren't impoverished. They weren't just trying to survive and get by. They were *thriving*. They were thriving by stealing from every citizen in the state of Florida and decimating essentially our cultural economy." Many locals didn't seem to see it that way. "I just love how the government can just mandate that I'm not allowed to do things" read one typical online comment after Operation Timucua. Ironically, when the state tried to protect cultural resources for the public, many members of the public saw it as yet another example of big government stepping on the little guy and his freedom. The total tally of artifacts lost in Florida will never be known.

Back at the Riverside Cafe, a waiter with a bandana wrapped pirate-style around his head came over to take our orders. The night before, I had seen the same waiter leap over the bar, catch a five-foot snake with a bucket, and wrestle it into submission. Now he was guiding us through a dizzying selection of sides—hush puppies, potato salad, coleslaw, fries, salad with a choice of six different dressings—that he reeled off in quick succession. At the next table, hoots and hollers rose up. A pair of alligators had appeared on the riverbank directly beneath the bar. People started tossing deep-fried oysters and scallops to the snapping jaws below.

Morgan Smith clarified that they didn't believe anything nefarious was going on in the Apalachee Bay. Each summer, divers fan out over the bay during the scalloping season, collecting the white shellfish from the seafloor, the scallops occasionally twittering up into the current to evade a diver's grasp, snapping like underwater castanets. The work requires divers to pick over the seabed closely, scanning the seabed the same way I saw Joy and Smith work. If looting had happened at any of their sites, the team said, it had probably happened opportunistically, not intentionally. Perhaps a scallop diver had pocketed a pretty projectile point, not knowing how much he had set back their team. I had done something similar when Smith handed me the flake on our first day of diving. I had been holding it up to the sun, admiring how smooth and thin it was. "Can I keep it?" I asked Smith, who winced. "No," he said, taking the flake back and slipping it into a plastic bag. "That would be illegal."

He and the other archaeologists don't want to alienate the good actors in the amateur archaeology community any more than Operation Timucua already did. Smith grew up an hour away in Tallahassee, where his parents still live. Walking around with him for even five minutes, you get a sense of how deep his roots in the Big Bend go. That guy Smith just greeted, yeah, he bought a boat off that guy. The Riverside Cafe we were sitting in right then, Smith's been eating here since he was a kid.

The hometown connections come in handy here. Local fishermen tell Smith about sites they might not share with outsiders. Amateur collectors will always discover the majority of sites in Florida simply because the nonexperts will always outnumber the experts in the field. The archaeologists need good actors to tell them about sites, or else they might never be discovered. "We need the public to be interested in archeology and preserving it," he said. "But we also don't want them to go out and do it themselves or see it as a source of income. If artifacts are stolen, looted, and sold, we will never know where it came from."

Our orders arrived in plastic baskets piled high with popcorn shrimp, deep-fried oysters and bay scallops, blackened grouper, lemon-peppered mullet. The sun started to set as we dug in, wiping our greasy fingers until the paper napkins were transparent, cracking beers open with satisfying pops.

For now, secrecy is perhaps the best protection an archaeologist can give a site, which is why Smith and Joy keep all their coordinates a secret. Archaeological sites on public land are ensured an extra level of confidentiality by Section 304 of the National Historic Preservation Act of 1966: the location of a historic resource is protected from freedom-of-information requests by the general public. Smith thought that the offshore sites in the Apalachee Bay were relatively safe just because they're so inaccessible. "I'd be surprised if some of those looters came here," he said of a site five miles offshore in deeper water. But Barbara Clark at FPAN saw it differently: "People who will never consider picking up an artifact off the ground, when they're diving, they see it completely differently and they're willing to take a piece off a shipwreck or something like that," she said. "I think it's just a different environment. There's nobody there really watching what you're doing."

One way to protect the sites would be to identify them as quickly as possible, using the HALD method. The oil and gas industry has decades of sonar data from sounding the continental shelf, with wavering haystacks buried within it that could pinpoint a vast array of still undiscovered offshore sites. "I can't believe that we've been collecting a lot of this data for decades and never put two and two together," Joy said. "We've been looking at the same HALD signatures for decades, and no one really knew what they were." He imagined artificial intelligence combing through old sub-bottom profiling data and then archaeologists ground-truthing each site with a dive. An uncrewed mapping drone such as the Surveyor might come in handy, and indeed, Saildrone recently opened an ocean-mapping headquarters in St. Petersburg,[44] right where so many submerged sites are turning up in Florida, including the seven-thousand-year-old burial site off Venice Beach.

6.

One or two scallop divers picking up artifacts from the seabed are probably not the biggest threat to undiscovered submerged sites. Sand mining is. "Down here, we do dredging and sand procurement," said Joy. Also known as beach replenishment, sand mining involves dredging sand from offshore sandbars and shoring up beaches or coastal real estate after hurricanes. All the marine life and artifacts found in that seabed are destroyed in the process.

At SEARCH, the cultural resources management firm where he works, Joy advises offshore industry on how to protect culturally sensitive seafloor. The federal government owns the continental shelf and leases parcels to companies to develop. Those companies are then required by the National Historic Preservation Act to hire consultants such as Joy to clear the sites for development. This is supposed to mean surveying for all types of archaeological sites, but the dominance and the visibility of shipwreck archaeology in the United States have meant that underwater precontact sites are often missed. "We've probably been destroying these sites for a long time," said Joy. He mentioned a beach in New Jersey approved for dredging by a team of shipwreck archaeologists in the 1990s. "As soon as they dredged it, they threw up a whole Archaic Period site onto the beach," he said. No one realized what had been destroyed until a woman walking on the beach afterward reported that there were about two hundred arrowheads scattered all over the sand.[45] A later investigation found that the survey had likely missed a valuable underwater site about a mile off the Jersey shore. The earliest destroyed offshore site I found dated back further, to the 1950s, when a company in Tampa Bay dredging for commercial oyster beds had turned up Paleo-Indian and Early Archaic Period diagnostic artifacts mixed in with an ancient shell midden. It had happened again and again around the low, half-submerged coast of Tampa Bay: at Apollo Beach, Turtlecrawl Point, and Terra Ceia Bay. Other sites off Galveston Beach in Texas and Long Island Sound between Connecticut and New York have also been destroyed over the years.[46]

Joy and his colleagues hope that the new HALD method will be a game changer in finding and protecting offshore sites. The Danish method used before was frustrating not only for archaeologists but also for developers, who would learn that the entire parcel of seafloor they had just leased from the government couldn't be developed because an ancient river ran through it. With development ramping up on North America's continental shelves, mapping the seafloor and identifying underwater sites need to happen *beforehand*—or we risk losing artifacts of a lost ancient world. We may even answer the beguiling question of how humans first came to the Americas.

Back at the Riverside Cafe, we polished off the last sips of warm beer and cold fries and said our good-byes. In just a few days' time, Memorial Day would arrive, along with the official start of hurricane season. The high-water marks of flooding by hurricanes past were scrawled all over the wooden beams of the Riverside Cafe, the way parents track children's heights on door frames: Dennis, 2005; Hermine, 2016; Michael, 2018.

Isn't sand mining a little counterproductive? I asked the archaeologists before we departed. Why move sand onto a beach, where the next storm will just wash it away again? Uh, yeah, the archaeologists agreed with a laugh. But that's Florida—got to protect the real estate. It's going to be a costly battle in the years to come. Florida is surrounded by water. It falls from the sky, up to 60 inches each year. It seeps up through the porous ground during high tides, a phenomenon known cheerfully as "sunny day flooding." And storms sweep in from the sea.[47] The ancient people of Florida knew that its lowlands provided little protection from the water and retreated. Will we learn to do the same? Or will we continue to dredge more sand in a Sisyphean struggle against the sea?

A few months later, I found myself flying back over Florida on my way to Kingston, Jamaica. From the air, I watched as a phalanx of dump trucks trundled along the Florida shore, moving sand out to the disappearing beaches. Beach replenishment is just one reason to mine the seafloor. Now I was on my way to an intergovernmental meeting in Jamaica to discuss another: deep-sea mining in international waters. Every few months, the governments of the world meet at the International Seabed Authority, a UN-associated agency based in Kingston, to discuss mining the seafloor beyond national borders. Mining proponents say that the precious metals are necessary for the world to transition to renewable energy. But in the deep sea, the most unexplored and uncharted terrain on the planet, we don't know what we might be destroying. As I looked down at the trucks working the Florida coast, the bright lights on land looked so small against the big black ocean.

CHAPTER 10

MINING THE DEEP

Give me a map; then let me see how much
is left for me to conquer in the world.

—Christopher Marlowe, *Tamburlaine the Great*, 1590

1.

In one of the first articles I ever read about Seabed 2030, ecologists warned that a finished map might open the seafloor to deep-sea mining.[1] It's a natural connection to make. Throughout history, intrepid explorers have descended on uncharted terrain, lugging astrolabes and theodolites, taking sightings, and collecting coordinates. Knowingly or not, they have laid the groundwork for the next inevitable stage: colonization, exploitation, and mass industrial development. "It is not that Lewis and Clark were men of conspicuous ill will," writes Stephen Hall in his book *Mapping the Next Millennium*, "it is that the domains that explorers chart, and the maps they produce, open up territories to interests that view them differently, interests that inevitably consume, exhaust, and extinguish the resources that are discovered, be they gold deposits or stands of timber or dispensable human culture."[2]

All the world's landmasses are now mapped and monitored by circling satellites, but where has this well-mapped world led us? Today, scientists are using satellites and maps to track the path of destruction: disappearing biodiversity and declining wilderness. A recent study surveyed the ocean's remaining wilderness and found that just 13.2 percent of the global ocean (over 21 million square miles) is still wild, with zero to little human impact. Much of this marine wilderness is found in the southern hemisphere, at extreme latitudes such as in the Southern Ocean, and in the deep, muddy-bottomed abyssal plains.[3] Perhaps if Seabed 2030 were left unfinished, the thinking goes, we might protect this last frontier from the capitalistic compunction to churn up more resources.

The deep sea is one of the most poorly understood ecosystems in the world, yet many of the top experts in the field already agree that the habitat will not bounce back quickly, if at all, from large-scale extraction of metals and minerals. "Given the very slow natural rates of recovery in most deep-sea ecosystems targeted for mining, loss of biodiversity in the deep sea is inevitable and may be considered to be 'forever' on human time scales," concluded one devastating report.[4] After a series of mining experiments in the Pacific during the 1970s and '80s, the tracks left by plows and rakes are still carved into the seafloor and very few animal populations have recovered to their original state.[5]

Deep-sea miners have spent decades prospecting in the Atlantic, Pacific, and Indian Oceans, but a tract of seafloor in the Pacific has drawn the most interest by far: the Clarion-Clipperton Zone (CCZ), a deep abyssal plain halfway between Hawaii and Mexico, that is roughly the size of Europe. In the most likely mining scenario, uncrewed nodule collectors, piloted from a ship above, would plow across this deep-sea prairie. The collectors are big, heavy machines[6] that resemble army tanks and run on caterpillar tracks, crushing all the life beneath them as they extract manganese nodules.[7] The CCZ is littered with nodules, clumped so densely in spots that the bottom of the ocean looks like the cobbled streets of London. Each nodule is rich in manganese and contains nickel, cobalt, copper, and rare-earth elements as well.[8]

This deep-sea treasure is a testament to the slow timeline of this underwater world. The nodules range in size from peas to potatoes, and if you were to slice one open, you would see layers like the rings of an ancient tree. These lumpy-looking black rocks are actually concretions: dense accumulations of minerals and metals scavenged from the ocean over millions of years.[9] At the core you might find a fragment of an ancient shell or the tooth of a shark, maybe even one from the greatest shark that ever lived: the giant megalodon, which swam through all the world's oceans 23 million years ago. The nodule builds up minerals around the hard center at a rate of 1 inch every million years.

As I write this in late 2022, commercial deep-sea mining has yet to begin, but that may change very soon. In June 2021, the president of the Pacific island state of Nauru triggered an obscure mechanism in the United Nations Convention on the Law of the Sea (UNCLOS) that allows a nation to fast-track and finalize mining regulations within two years.[10] After decades of debate at the International Seabed Authority (ISA) in Kingston, Jamaica, the intergovernmental negotiations were aiming to finish the Mining Code by June 2023,[11] a timeline that many observers saw as rushed, if not impossible in the already strained proceedings. At the last ISA meeting before the COVID-19 pandemic, the talks grew so tense that the African Group, representing forty-seven countries, threatened to walk out if one of its long-standing concerns, that of the geographic diversity on a key decision-making committee, was not addressed. In the world of diplomacy, this is the equivalent of shots fired.

So will Seabed 2030 become a treasure map to the world's last frontier? The deep-sea ecologists opposed to Seabed 2030 were also arguing against a scientific tradition that always demands more data, more knowledge, and ultimately more management. Yet when it comes to protecting the planet, studies have shown that nature may recover best when it's left alone or, better yet, never developed at all. Wilderness refuges maintain functioning, interconnected ecosystems, animals found nowhere else on Earth, and species with higher genetic diversity—all of which might make them more resilient to climate change than restored or managed ecosystems.[12] Old-growth forests, for instance, absorb and store carbon better than do replanted monoculture plantations, whereas second-growth forests sequester carbon at a faster rate when they're restored naturally.[13] The deep sea, uncharted, unknown, could be the world's largest intact wilderness refuge, as well as a bulwark against runaway climate change transforming the planet from top to bottom.

It was an alluring dream, but when I wandered into the ISA library, I realized just how futile that dream was. The library was full of seafloor maps. State-backed mining corporations had already mapped and staked claims to the richest tracts of seafloor. The resolution of Seabed 2030 would be far too grainy to pinpoint new mineral deposits on the seafloor. It was the miners, not the academic mappers at Seabed 2030, who had better maps. I pulled down an oversized atlas from a random file cabinet and opened it to a page displaying seamounts in the South Pacific. Each page detailed a new seamount, mapped in the three-dimensional rainbow bathymetry that had become familiar. Alongside each map was a legend that broke down the seamount into its mineral composition, with percentage yields for manganese, iron, cobalt, nickel, and copper. When the map was made by a Japanese survey vessel in the 1980s,[14] there was a glut of metals on the world market and most ocean-mining projects were on hold.[15] Forty years later, the metals outlook has changed dramatically. The demand for nickel, cobalt, iron, and copper is surging thanks to the rise, ironically, of carbon-saving gadgets such as electric vehicles, wind turbines, and solar panels.[16] With the treasure maps already drawn, the government delegates downstairs just might offer the deep sea its best protection against this experimental new industry.

Over the last twenty-five years, government delegates have met at the International Seabed Authority to negotiate regulations for mining the deep sea, but never with the urgency of today. Thanks to the obscurity and magnitude of the deep sea, most people have no idea that the negotiations are happening. "The public hasn't weighed in 'cause the vast majority don't know what deep-sea mining is," the oceanographer Jeffrey Drazen told Alan Jamieson on *The Deep-Sea Podcast*.[17] According to Drazen, one mining contractor will have to mine 115 to 230 square miles of seafloor each year to turn a profit.[18] There are seventeen contractors prospecting in the CCZ alone, and each contract period lasts fifteen years. We are about to commence the single greatest human alteration to the surface of the planet, and we will very likely end up damaging the seafloor before we finish exploring it.

2.

All week long at the 26th Session of the International Seabed Authority, I sat next to Arlo Hemphill, Greenpeace USA's senior oceans campaigner. My first impression was that he seemed a bit on edge. He tapped on his laptop at a furious pace, barely looking up at the negotiations unfolding in front of us. Before he switched his phone to vibrate, the ring tone was a police siren. But I would be on edge, too, playing the confrontational role that Greenpeace does at such international meetings. "Greenpeace attends this 26th Session of the ISA *extremely concerned* about the way things are evolving within this international organization," Hemphill told the gathered delegates early in the proceedings. Greenpeace was calling for nothing short of a moratorium on mining the deep sea,[19] but at that point no country had joined that call; not even Costa Rica, which was pushing ardently for environmental protection, seemed ready to say the M-word just yet.* Later that week, I would find out that Hemphill had a few more reasons to feel nervous.

With fewer than a hundred people in attendance due to Jamaica's COVID-19 restrictions, the 26th Session felt a little sparse. The wood-paneled assembly room at the Jamaica Convention Centre recalled the

---

* In 2022, Chile became the first country to support a fifteen-year moratorium on mining the international seafloor. Since then, Palau, Fiji, Samoa, France, Spain, Chile, New Zealand, Canada, Germany, and Costa Rica have all come out in support of a ban, a moratorium, or a "precautionary pause" on deep-sea mining until more is known about its impact on marine life. See "Chile Calls for a Moratorium on Deep-Sea Mining," Deep Sea Conservation Coalition, June 20, 2022, https://savethehighseas.org/2022/06/20 /chile-calls-for-a-moratorium-on-deep-sea-mining/; Elizabeth Claire Alberts, "A Year Before Deep-Sea Mining Could Begin, Calls for a Moratorium Build," *Mongabay*, June 30, 2022, https://news.mongabay.com/2022/06/a-year-before-deep-sea-mining -could-begin-calls-for-a-moratorium-build/; Karen McVeigh, "Row Erupts Over Deep-Sea Mining as World Races to Finalise Vital Regulations," *Guardian*, March 21, 2023, https://www.theguardian.com/environment/2023/mar/21/row-erupts-over-deep-sea -mining-as-world-races-to-finalise-vital-regulations; and "Canada declares moratorium on deep-sea mining at global ocean conservation summit," Canada's National Observer, February 9, 2023, https://www.nationalobserver.com/2023/02/09/news/canada -declares-moratorium-deep-sea-mining-global-conservation-summit.

grandeur of the UN General Assembly in New York: the raised rostrum at the center of the room, the desks fanned outward in order of the attendees' importance. I took my seat at the back with the other observers from Greenpeace, the Pew Charitable Trusts, and the Deep-Ocean Stewardship Initiative. There's only so much you can influence from the sidelines, the observer from Pew told me. The most important thing is to get your objection onto the record.

As the council president gaveled in the meeting and opened the 26th Session of the authority, the atttendees fit the translator headpiece over their ears and spun the dial under their seat to find the correct language. Spanish and French are spoken, but English is the lingua franca, a very formal English with lots of "sirs" and "madams" and legalese such as "In keeping with the provisions of article six of annex three to the United Nations Convention on the Law of the Sea, paragraph three, regulation 21" that would send me scrambling to the ISA website to download the correct document. By the time I had scanned the six- or ten-page document, the negotiations had long since moved on.

The ISA secretary-general, Michael Lodge, sat at the center of the raised rostrum. Throughout his tenure at the ISA, Lodge has taken a pro-mining stance, appearing in promotional videos for a deep-sea mining company wearing a hard hat with the company's logo emblazoned across the front.[20] In his opening remarks, Lodge urged the ISA Council to grant nations the right to mine the seabed and said that otherwise the ISA would be failing to fulfill its mandate as the overseer of the deep-sea mining industry. He gave only brief mention to the second half of the ISA's dual mandate. As set out by UNCLOS, the ISA's mandate is to develop the resources of the international seabed for the benefit of all humanity *and* to ensure that that development never harms the environment.[21] However, as commercial deep-sea mining edges closer to reality, "More people and more governments are questioning whether it is a good idea to have an organization that promotes deep-sea development also write the rules that are meant to protect its environment,"[22] wrote the authors of a 2018 International Union for Conservation of Nature (IUCN) report on deep-sea mining. Critics have also pointed out that the ISA embodies a glaring conflict of interest: it is a mining regulator dependent on mining royalties to operate.[23]

As the delegates got down to business, it quickly became clear that the session would deal very little with the regulatory meat of mining the seabed or indeed mention the seabed at all. Inside the air-conditioned chamber filled with lawyers and policy experts tapping at computers and cell phones, the negotiations unfolded in a predictable fashion. When a delegate wished to speak, she stuck her country's placard up at a ninety-degree angle into the slot in front of her. The council president worked around the room, granting delegates a chance to speak in turn: "Distinguished delegate of Germany, please go ahead." A red light on the delegate's microphone switched on, the delegate read a set of prepared remarks, and the discussion moved on. The speeches generally lasted from two to five minutes and followed the "shit sandwich" format. The delegate would begin with some pro forma niceties, such as congratulating the secretary-general on his reappointment or thanking Jamaica for hosting the meeting. Then would come the shit: a pointed observation, criticism, or suggestion that showed a country's hand. Then the delegate would finish with a congratulation or compliment. After the last three years of living through the chaos of a global pandemic when public health rules had been openly flouted, I enjoyed a bit of rule following these days. But the council felt almost comically divorced from the rough reality of working offshore.

There are few people to observe and report crimes at sea, and so more crimes are committed with impunity. "The rule of law—often so solid on land, bolstered and clarified by centuries of careful wordsmithing, hard-fought jurisdictional lines, and robust enforcement regimes—is fluid at sea, if it's to be found at all," writes the ocean journalist Ian Urbina.[24] Fertilizer runoff bleeds into the sea. Garbage patches filled with plastic pollution churn far from land. Fishing trawlers rake the seafloor, destroying habitats and overharvesting depleted stocks. Fisheries observers, who ensure that fishers follow regulations, go missing or die in mysterious ways.[25] A new experimental industry entering this murky offshore world is an ominous harbinger for the marine environment. Deep-sea mining will take place in darkness, far from land, with very few observers on hand. Remotely operated vehicles such as *Hercules* could be deployed to monitor the damage, but they probably won't be, or at least not at the scale needed. The proposed mining sites are vast, and ROVs are prohibitively expensive and technically challenging to operate. On the first day of the meeting, the delegate from Chile proposed that mining operators should be audited to ensure that they follow environmental regulations—a good idea I never heard raised again.

When Hemphill spoke at the ISA, his words fell like a bludgeon compared to the diplomatic cadences of the delegates. Near the end of the second day, which had been almost entirely devoted to discussing a work schedule for meeting the two-year deadline triggered by Nauru's proposal, the floor opened at last to observers, who were allowed to read a statement, known as an intervention. "With the triggering of the two-year rule by Nauru last June," Hemphill read in Greenpeace's intervention, "there is a real threat that the ISA will rush into deep-sea mining in spite of its legal obligation, first and foremost, to ensure the protection of the marine environment, including the seabed, which is the common heritage of humankind."[26]

Hemphill didn't come up with the lofty phrase "the common heritage of humankind" himself. It comes from a famous speech that laid the groundwork for UNCLOS and is embedded within the international treaty governing the high seas. In 1967, Arvid Pardo, the Maltese representative to the United Nations, stood up on the floor of the UN General Assembly in New York and delivered a three-hour filibuster of a speech on the resources of the global ocean.[27] He spoke of the growing militarization of the seabed and the stationing of nuclear weapons on seamounts,[28] detailed the rise of aquaculture and of scientists mastering fish husbandry to feed the world, and referenced a now-outdated paper estimating that more than a trillion tons of precious metals might be lying on the seabed.[29] His point was that all those resources could no longer be left to the freedom of the seas, a legal concept developed by a Dutch jurist in the seventeenth century.[30] If the status quo were to continue, he predicted, the rich would get richer and the poor would get poorer. Another race for resources was afoot, Pardo warned, one that echoed the colonial scramble for Africa at the turn of the previous century, only this time at the bottom of the ocean.[31]

If the United Nations acted now, Pardo said in his 1967 filibuster, and gave the international seafloor legal status as the "common heritage of mankind," the international agency could regulate a resource *before* extraction began. "I thought it could serve as sort of a bridge to the future and unite the world community in its quest to preserve our planet for generations to come," Pardo told interviewers years later.[32] Rich nations, with the money, science, and technology to mine the seafloor, were less enthusiastic about Pardo's proposal.[33] But the speech struck a chord with a number of developing nations in Africa, many of which had recently achieved independence, as well as the Soviet Union and Eastern European countries, which liked the socialistic approach rather than the capitalistic free-for-all favored by the West.[34]

Pardo's speech sparked another ten years of negotiating over what would eventually become UNCLOS.[35] When the delegates sat down to negotiate managing the international ocean, they based their positions on the most up-to-date science of the day, but when it came to life on the seafloor, that science was anything but settled.

For centuries, experts debated whether life could even exist at the bottom of the ocean. During the Victorian era, ocean experts hewed to the popular azoic theory, which held that nothing could live beyond 300 fathoms (about 550 meters, or 1,800 feet) deep.[36] It took years for naturalists to accept that fishermen and whalers were pulling up life far beyond 300 fathoms.[37] In 1960, two men reached the bottom of the Mariana Trench for the first time inside an awkward-looking submersible called a bathyscaphe. As they touched down at 10,916 meters (35,814 feet) deep, one reported seeing a flatfish flicker past the porthole. In the media flurry that surrounded the historic dive, the sighting became proof that life could exist at full ocean depth. Just a few years later, experts called that sighting into question; today, most biologists agree that it was probably not a fish at all, thanks to a new hypothesis that fish might be physiologically incapable of surviving beyond 8,500 meters (27,887 feet) deep.[38]

Then, in 1977, Bob Ballard and company found a lush community of deep-sea animals clustered around a Pacific hydrothermal vent off the Galápagos—and named it the Garden of Eden. That discovery, now recognized as one of the greatest of the twentieth century, upended everything we knew about life on Earth.[39] An entirely new chemosynthetic world came into view, a parallel underwater existence that required no sunlight at all. "None of this was in the imagination when UNCLOS was negotiated" during the 1970s, Philip Steinberg, a geography professor and the director of IBRU, Durham University's Centre for Borders Research, told me.

There are three regions in the deep sea that miners plan to target: the seamounts, the abyssal plains, and the hydrothermal vents along ridges and fracture zones. In the case of the seamounts and hydrothermal vents, deep-sea miners would grind down the vents and shave off the tops of seamounts, both of which are crusted with minerals. In the plains, miners would send their crawling tanks to either cut into the seafloor or vacuum up manganese nodules that lie exposed and loose on the muddy seafloor like the potatoes they are so often compared with.[*]

Ever since the Garden of Eden was discovered, deep-sea scientists have continued to uncover more evidence that all three regions are home to rich and diverse marine life. Hydrothermal vents gush chemical-rich water that sustains tube worms, shrimp, and all manner of unknown species. These vents, or ones similar to them, very likely gave birth to the first life on Earth. Seamounts are biodiversity hot spots, serving as feeding and spawning grounds far from land.[40] Manganese nodules are the hard foothold on which the ecosystem of the deep-sea prairies is built.[41]

By the time UNCLOS was opened for ratification in 1982, Arvid Pardo, who would become known as the father of the treaty, had grown disillusioned with his offspring.[42] "It is probably the most inequitable treaty that has ever been signed in the world," he said with disgust in 1981. The United States had played a key role in drafting UNCLOS, and Pardo thought that the treaty favored rich nations by granting them the most accessible underwater sites.[43] At the last minute, the United States pulled out of signing the convention, due largely

---

[*] In the mining industry, the metal deposits on hydrothermal vents are referred to as either "seafloor massive sulfides" or "polymetallic sulfides," and those on seamounts as "cobalt-rich ferromanganese crusts" or "polymetallic crusts."

to the "socialistic" wording of "the common heritage of mankind" and what that phrase might mean for deep-sea mining if, or when, it ever happened. The United States also convinced a number of other nations to do the same.[44] UNCLOS languished in regulatory limbo until 1994, when, after many more edits to appease industrialized nations, it gained enough signatures to be ratified. The United States again declined to sign and has not signed to this day.[45]

Somehow Pardo's original phrasing made it through the final round of revisions, and the seafloor's legal standing as an international common, owned by no one and managed by the ISA, remains in place. "The common heritage of mankind" has now become a legal template for other shared geographies: Antarctica, the moon, even the human genome. But what exactly "the common heritage of mankind" means for deep-sea mining depends on whom you ask. "What the western and northern states meant by it is 'Whoever gets there first gets it.' And what the global South meant was 'It belongs to all of us, and whoever gets there first has to share,'" said Aletta Mondré, a professor of political science at the University of Kiel. "What it actually means in practice, we don't know."

3.

At critical junctures in history, blank spaces appear on the map, ripe for exploring and exploiting.[46] If you ask the people at the Metals Company what the Pacific's Clarion-Clipperton Zone is like, they'll likely tell you that the CCZ is another blank spot on the map. "It's a pretty barren landscape down there. It's the biggest desert on the planet; it just happens to be underwater," the company's CEO, Gerard Barron, has said, in different formulations, numerous times in interviews.[47] The Metals Company is behind three of the twenty-two mining contractors prospecting on the international seafloor today. Barron often carries a manganese nodule with him to media interviews, which he declares to be "a battery in a rock."[48] To me, the choice Barron seems to be offering is clear: if we want to save the planet, if we want to save *ourselves*, all we have to do is sacrifice the barren desert at the bottom of the ocean.

Perhaps the most famous example of a blank spot on a map spurring exploration and exploitation is the enduring myth of an empty America before European colonization. On early maps, the Americas are often drawn as an uninhabited wilderness despite the 50 million to 70 million people who lived there at the time Columbus made landfall in the Caribbean.[49] Another famous example comes from *Heart of Darkness*, Joseph Conrad's novella, written at the height of Great Britain's imperial might: "At that time there were many blank spaces on the earth, and when I saw one that looked particularly inviting on a map (but they all look like that) I would put my finger on it and say, When I grow up I will go there," says Marlow, the doomed colonialist bound for a terrifying journey up the Congo River.

When I was a kid, I used to run a finger over a spinning globe, too, imagining what life was like in faraway places—a game I remember as innocent and playful. When you're a kid, too short to see over countertops and tables, maps provide an exhilarating, godlike perspective on the whole world. But the map historian J. Brian Harley sees something darker in Marlow's attraction to blank spaces. "In this view," he writes, "the world is full of empty spaces that are ready for the taking by Englishmen."[50] Herein lies the conflict between exploration and exploitation that is bound up in any map of a new frontier. The blank spaces are never truly empty, but in the vacuum of firsthand experience, people fill them with stories that best serve their ambitions.

The CCZ in the Pacific Ocean does appear a little blank at first: an empty landscape you might expect to see on Mars or the moon. However, scientists are discovering that it is home to remarkable biodiversity. Patricia Esquete, a deep-sea biologist who sat next to me during the ISA negotiations in Jamaica, has gone on three research cruises to the CCZ, where she studies the marine macrofauna of the deep, mostly tiny invertebrates that can be seen with the naked eye, such as clams, snails, shrimp, worms, and crustaceans. To understand the ecosystem, she lowers a steel box, known as a box corer, down to the bottom of the ocean, where it punches out a neat square of seabed. Then she reels it back in, all the animals in it preserved in situ.

Scientists believe that thousands of species might be unique to the CCZ. Many of the animals that Esquete has encountered are new to science or undescribed, and all have been taken from an area that represents less than 1 percent of the entire mining site. Nearly half of the larger animals there depend on the manganese nodules for a sturdy habitat in the otherwise soft sediment.[51] The Casper octopus, a ghostly white wisp of a creature discovered in 2016, lays its eggs on sponges that grow on the nodules.[52] Inside the cracks and holes of the nodules live even smaller animals, such as nematode worms[53] and tardigrades, also known as "water bears" for their otherworldly and adorable appearance. "Basically, if you get rid of the nodules, the ecosystem disappears," Esquete explained. And, she stressed, this is not only about damaging an unknown ecosystem. "Even if you did take the ethics away from the conversation, there are still good reasons not to mine."

Sediment plumes are one of the biggest environmental concerns surrounding deep-sea mining. The plumes, kicked up by nodule collectors scraping the seafloor and by tailings pumped back down into the ocean, may travel hundreds of miles from the mining site[54] (deep-sea miners say as few as six miles).[55] There are no physical borders at sea to stop the swirling plumes. Aboard E/V *Nautilus*, I watched ROV *Hercules* zip over the powder-fine sediment at the bottom of the Santa Barbara basin, sending up a trail of dust. That made it easy to picture the underwater dust storm that a 50-foot-long,[56] 25-ton collector[57] would stir up as it vacuumed up manganese nodules from the seafloor like a deep-sea Roomba. (Another design of nodule collector would cut a few inches into the seafloor with a row of metal teeth, extracting the nodules as it goes.) The slurry would be pumped up to a surface vessel, the nodules plucked out, and the unwanted dregs pumped back down a discharge pipe and released at the mining site or higher up in the water column, perhaps at around 330 feet deep.[58] During the process, the crushed ore would release metals, potentially toxic ones, into the sediment that is dumped back into the ocean. Wherever the dust came to settle, it would very likely harm the deep-sea animals unlucky enough to be living there. The sediment would drag down delicate deep-sea octopuses and gossamer worms drifting in the water column. It would clog the filter-feeding appendages of sponges, jellyfish, and clams. It would absorb the light that bioluminescent animals use to feed and communicate.[59] Whale sharks, leatherback turtles, and seabirds also pass right though the planned mega mining site of the CCZ. The health impacts on a turtle or tuna swimming through a dust storm tainted by heavy metals are still unknown. The larger animals naturally feast on the smaller ones, which could pass mining waste up the food chain into larger animals and humans as well.[60] Huge herds of migrating tuna swim around the CCZ, where half the world's tuna stocks originate.[61]

When I try to imagine a human translation for the sediment plumes that deep-sea mining might unleash in the CCZ, I picture farming communities during the Dust Bowl of the 1930s. Caused by aggressive farming practices that had stripped off the prairie topsoil, apocalyptic dust storms sucked the life out of agricultural towns from Oklahoma to Saskatchewan.[62] The fine silt—thick with silica, a component of glass—killed people and livestock, destroyed crops and livelihoods, and set off the great "Okie" migration to California. Those towering dust storms, as tall and wide as mountains, advanced across the prairie, blotting out the sun and plunging towns into darkness for hours, sometimes days. Many survivors recalled thinking that the dust storm was a sign of impending Armageddon—and for many it was, leaving devastation in its wake.[63] Undersea sediment plumes might be even worse than those dust storms. The oceanographer Jeffrey Drazen has found that sediment plumes could muddy the waters for several *years*: "Indeed, the concern is greater for the ocean in which suspended particles settle much more slowly in seawater than in air."[64] Meanwhile, the nodules that form the bedrock of this fragile community are stripped away.

Industrial machinery will also bring light and noise into an ecosystem that has been dark and silent for billions of years. The blinking lights of the underwater machinery might confuse deep-sea animals used to life in the dark. The noise of the nodules rattling up the pipes connecting the mining tanks to the surface ship would likely reverberate through the silent water, disorienting the sightless creatures that use sound to navigate and hunt.[65]

The Metals Company often positions itself as the climate-friendly alternative to the oil and gas industry: "I think we absolutely have to drive towards *ending extractive industry for good*," Gerard Barron told a clean-tech magazine in 2021. ". . . It's like [the nongovernmental organizations that criticize deep-sea mining] are ignoring what the biggest threat is and that's global warming and [CO2] emissions."[66] Such claims fall apart when the potential climate risks posed by mining the deep sea are considered. Since the Industrial Age, the global ocean has sequestered the majority of the heat from climate change. (NASA recently reported that 2022 was the ocean's warmest year on record.) The ocean is already struggling to handle the heat, along with an onslaught of other human-caused pressures such as acidification, deoxygenation, pollution, and overfishing.[67] The deep sea plays an important role as a carbon sink where millenniums of dead animals sink down and are entombed in sediment. It is also home to microbial communities that play a little-understood role in cycling carbon through the ecosystem.[68] If large tracts of seafloor are mined over long periods of time, there could be serious implications for mitigating climate change as well.[69] In 2022, thirty ocean experts published a paper voicing their concerns about the unknown impacts of deep-sea mining, noting that changing climate scenarios and extensive mining operations could make the damage exponentially severe.[70]

Simply put, if we were to lose or even diminish the ocean's ability to sequester or cycle carbon at this delicate moment in time, when the international community is already struggling to lower its carbon emissions, we may risk tipping the scales toward climatic chaos.

Rather than ending resource extraction and sustaining the circular economy, deep-sea mining appears to swap one form of extraction for another. The Metals Company, along with its Swiss partner Allseas, is in the midst of converting a mega drill ship once owned by the Brazilian oil corporation Petrobras to a nodule collector.[71] Some of the same technology and operating procedures developed by the offshore oil and gas industry will simply be transferred to deep-sea mining.[72] The mining ships will run on heavy fuel oil during their long transits to and from remote sites and during heavy extraction periods in the middle of the Pacific.[73] No one can say exactly how many emissions deep-sea mining might produce because commercial extraction hasn't begun yet, but one hypothetical operation in the Pacific producing 3 million dry tons of manganese nodules could emit up to 482,000 tons of CO_2,[74] equivalent to the emissions produced by the energy use of 55,079 American homes in one year.[75]

The profits to be gained from this experimental new industry do not seem all that worth it. In 2020, Fauna & Flora International pegged deep-sea mining's worth at around $2 billion a year, yet it could disrupt the far more critical small-scale fisheries on nearby Pacific islands. Artisanal fishing is a major employer in small island states, where it also provides up to 90 percent of the protein consumed there. Fishing is also an irreplaceable part of Pacific islander culture. In a string of coastal villages in Papua New Guinea, the locals practice "shark calling": singing to sharks before hunting them. Elders there blame, in part, a deep-sea mining operation trialing equipment in their waters with scaring away the sharks and threatening a custom that binds their people to the sea.[76]

Mining also threatens to derail deep-sea industries that have yet to truly take off, such as the marine biotechnology market, which is estimated to reach $6.4 billion in value by 2025. One of the most famous and profitable antiviral drugs ever created comes from two compounds found inside the sea sponge *Cryptotethia crypta*, released as the first HIV therapy in 1987.[77] In San Diego, where I live, biotechnology companies are sprouting like weeds along Interstate 5 running north to Los Angeles. So far, most of the marine compounds for new pharmaceuticals have come from corals and sponges, which are easiest to collect. In 2020, more than a half dozen drugs taken from marine products had already been approved, and another twenty-eight were in clinical trials,[78] with more and more being developed all the time. (Drug companies invest in producing a new drug once the marine compounds can be made synthetically—no wild animals are necessary for its mass production.)[79] One compound from a deep-sea hexactinellid sponge could help in the fight against the virulent superbug methicillin-resistant *Staphylococcus aureus*, or MRSA.[80] The World Health Organization has called antimicrobial resistance one of the top ten global public health threats, and considering the recent COVID-19 pandemic and the surge of antibiotic-resistant superbugs, the world very much needs new drugs to fight them.[81] The biotech industry is experimenting with a raft of new marine-derived products for addressing global challenges, including using red algae to reduce methane emissions from livestock and fortifying staple food crops with added nutrients.[82]

From undiscovered species and bustling biodiversity to the economic and cultural value of maritime activities new and old, it's clear that the abyssal plains are not the barren desert that the Metals Company would have us believe. But perhaps what matters most is that the ocean floor often *appears* empty—and world maps have trained us to see the ocean that way. Colored a flat blue on traditional world maps, the eye skips over the ocean, the biggest "blank space" on Earth, on its way to land. This bolsters the myth that the ocean floor is empty, lifeless, and not worth exploring.

Deep-sea mining advocates often wave away the concerns of scientists such as Patricia Esquete who want to pause extraction and collect more data on the deep-sea habitat first. Scientists *always* want more data, they say; that's the nature of science. So when will we know enough about the deep sea to start mining? And who gets to decide when is enough? Esquete couldn't answer those questions when I asked her, but she told me that the science is already clear on one thing: "A lot will be lost." What she is most passionately opposed to, the reason she traveled all the way from Portugal to Jamaica to read out statement after statement condemning seabed mining at the International Seabed Authority meeting, is the two-year deadline set by Nauru. That, she said, is definitely too soon.

4.

On the last day of the meeting, the negotiations closed to observers. As delegates hammered out a work schedule to meet the two-year deadline behind closed doors, Esquete and I wandered outside the ISA compound to grab lunch out on the streets of downtown Kingston. Throughout the meeting, the attendees had been mostly shielded from hardship in the host country. Police escorts shuttled delegates to and from hotels. Guards lined the corridors of the conference center. But as soon as we stepped outside the compound walls, the rough reality of everyday life in Jamaica flooded in. The pavement was riven with potholes the size of craters, and the smell of sewage and smoke wafted over the street. We passed a man fishing from the Kingston pier with a line wrapped around an empty plastic bottle for a rod. When we asked to see his catch, he showed us a few small, silvery fish that he planned to grind into soup for dinner. We passed a dying dog with a bloody, flattened paw lying in the shade, heaving its final breaths. All that in a single block in the downtown harbor area of Kingston, lined with gleaming high-rises. When we reached the café, we ordered lunch from a girl who looked as though she should be in middle school.

Arvid Pardo's dream was that one day, seafloor mining would level the playing field between rich and poor. Instead, the ISA felt like a rich island barricaded against a sea of poverty outside. Back in the 1980s, Jamaica won its bid to host the ISA by arguing that an agency devoted to social equity should be based in a developing country. But the wealth of an international agency barely seemed to make it past the security perimeter. The ongoing negotiations over the international seafloor have become an unlikely flashpoint in the struggle between the rich and the poor, the global North and the global South.[83]

Of the three types of seafloor terrain that miners plan to target, the best environmental argument can be made for manganese nodules. Hydrothermal vents would be mined primarily for zinc and gold, which are not in huge demand for the fight against climate change.[84] Seamounts are not considered all that valuable.[85] Nodules, however, lie loose on the seafloor and are rich in the nickel and cobalt that are needed for renewable tech and are so problematic to mine on land. This is another argument for deep-sea mining: that it would help end destructive mining practices in developing countries. Much of the world's cobalt, for example, is extracted in the Democratic Republic of the Congo, where the industry is poorly regulated and child laborers work in horrendous conditions, while nickel mines in the Philippines,[86] Indonesia,[87] and New Caledonia[88] mow down tropical rain forests and pollute waterways. The latest trend among electric automakers, however, is to move away from cobalt and nickel and toward lithium-iron phosphate and other battery alternatives.[89] Deep-sea mining provides no guarantee that mining on land would end if this new offshore industry began. The two would most likely continue in parallel, albeit with diminishing returns for impoverished countries whose main source of revenue is mining. That concern is apparent among the mining-dependent countries across South America and Africa, which seem the most opposed to deep-sea mining at the ISA.[90] The African Group ran the numbers on just how much member nations might earn from the ISA's still-undecided royalty system, which is supposed to compensate primary resource–heavy economies for their losses. The report found that a typical nation would receive a paltry payout of less than $100,000 per year.[91] "The African Group does not consider that this is fair compensation to mankind," the Algerian delegate told the ISA meeting in February 2019.[92]

Thanks to the "common heritage" principle baked into UNCLOS, nations, not corporations, are supposed to profit from deep-sea mining, but over time, a sophisticated network of shell companies, subcontractors, and partnerships based in the global North has managed to contract for control of the richest parts of the seafloor. Just four entities, all linked to wealthy industrialized nations, dominate the mining sites in the CCZ. At an ISA Council meeting in 2019, Gerard Barron, the CEO of the Metals Company, the deep-sea mining company that appears to be furthest ahead in the race to mine the CCZ, took Nauru's microphone and spoke on behalf of its sponsoring country. "Personally, I get very uncomfortable when people describe us as deep-sea miners," he told the room.[93] (The company has called its operations "harvesting.") Barron is an Australian running a company based in Canada, where many of the company's shareholders live or have strong ties to the global North. Rather than redistributing the riches of the seafloor to developing nations, as Arvid Pardo envisioned, a *New York Times* investigation revealed that the ISA had shared confidential data with executives at the Metals Company that had allowed the corporation to seize the most profitable areas of seafloor first and find developing countries to back their bid later. (The ISA denied "improperly or unlawfully" sharing confidential data.) A Tonga community leader told the *New York Times* that in exchange for sponsoring the Metals Company, the country would receive less than half a percent of the company's total estimated value of mined material.[94]

If you've never heard of the Metals Company before, that's understandable. Before September 2021, it didn't exist. Shortly after the president of Nauru pulled the two-year trigger at the ISA in summer 2021,[95] a precursor company called DeepGreen Metals merged with a special purpose acquisition company—a lightly regulated Wall Street invention modeled after the fraudulent blank-check companies of the 1980s—and went public as the Metals Company.[96] Today, the Metals Company is expected to become the first commercial deep-sea mining company in the international seabed. But a decade earlier, Nautilus Minerals was in a roughly similar position and had some of the same key backers as DeepGreen and the Metals Company, including Barron, an early investor in the forerunner company.[97] Nautilus acquired an ISA permit to prospect in the CCZ on behalf of Tonga and Nauru—permits that the Metals Company now holds—but Nautilus also struck a deal with Papua New Guinea to mine hydrothermal vents inside its territorial waters. Because the ISA's jurisdiction extends only to international waters, Nautilus could move ahead without the finalized Mining Code.[98]

There are about six hundred known hydrothermal vent fields around the world that erupt along the midocean ridge system and in fracture and subduction zones. These hot spots are rich, self-contained ecosystems and most are about a third the size of a football field.[99] The Papua New Guinea site seemed to contain deposits sufficient for only two years of mining—nowhere near the fifteen years needed to turn a profit.[100] The idea was for Nautilus Minerals to jackhammer apart the hydrothermal vents and then pump the slurry up and back down, similar to the manganese nodule mining that the Metals Company plans to do in the CCZ.[101] One of the chimneys off Papua New Guinea's coast sits inside a glass case right off the map library at the ISA headquarters. Donated by Nautilus Minerals, the black smoker, which would look far more at home in Sauron's dark tower from *The Lord of the Rings*, sits next to a picture of a robotic arm cracking the chimney from the seafloor.

In 2018, Nautilus defaulted on a payment for a new ship and filed for bankruptcy protection the following year. The government of Papua New Guinea had borrowed heavily to purchase a major stake in the project, and the country, along with its taxpayers, lost upward of $120 million.[102] Officials from Papua New Guinea now call the project a failed investment, and the country has joined other Pacific islands opposed to deep-sea mining.[103] Long before Nautilus's money problems, CEO David Heydon and Gerard Barron had stepped away from the company, profited handsomely, and moved on to starting DeepGreen. In Barron's case, he "turned a $226,000 investment into $31 million, and he successfully exited his position near the height of the market," according to a mining industry magazine.[104]

During the ISA Council meeting, an environmental lawyer with the Deep Sea Conservation Coalition asked who *really* controlled the subsidiary companies mining the seafloor. "Where is the effective control of NORI, a subsidiary of the Metals Company? If not in Nauru, where is it? And what are the consequences of that?" asked Duncan Currie. "This effective control issue previously arose on the acquisition of NORI from its previous owner, Nautilus, which went into liquidation, at the cost to Papua New Guinea of over $100 million."[105]

Throughout the negotiations, a representative from the Metals Company sat behind the Tonga delegation but never addressed the council. I watched him mix freely with the delegates, who referred to him by either first name or "that guy who looks like Anderson Cooper." When I approached to ask for comment, he said that he would have to see my questions in advance. As a journalist, I generally try to avoid that situation, and interviewees rarely ask, but I obliged and sent my questions. He still declined to speak.

A delegate from Micronesia told me that Pacific islanders are generally united on issues, but deep-sea mining is where they divide. Micronesia, Fiji, Palau, Tuvalu, Samoa, and Guam are opposed to it; Tonga, Nauru, Kiribati, and the Cook Islands have all backed subsidiary mining companies. Kiribati is one of the least developed countries in the world, its future severely threatened by rising sea levels.[106] In 2022, an underwater volcano near Tonga erupted, and the ensuing tsunami and ashfall caused $90 million in damages—roughly a fifth of the country's GDP. The Pacific island community is still reeling from COVID-19 after many countries closed their borders and restricted international tourism. In 2020, Fiji's economy contracted by 19 percent due to the precipitous drop in tourism. With financial losses piling up, a risky new venture becomes all the more appealing to a poor nation with few other options.[107]

Nauru's government has a habit of making bad business decisions. The island's primary resource was guano—that is, poop left by seabirds that use the island as a rest stop during long transpacific flights. For

much of the twentieth century, mining and exporting guano for fertilizer made Nauruans rich. It also reduced the 8-mile-square island to
a strip mine, picked over and hollowed out by extraction. The government took the guano money and invested it in a series of failed busi-
nesses overseas: casinos, hotels, even a London musical that flopped.[108]
If Nauru's gamble with deep-sea mining pans out, the country could
receive a reported $100 million every year.[109] Of course, there's also the
chance that the Metals Company and Nauru will succumb to the same
fate as Nautilus and Papua New Guinea before it. During the Metals
Company's public offering in 2021, which was supposed to bring in
hundreds of millions of dollars, prospective investors questioned just
how sustainable deep-sea mining truly was. The offering was a bust.
The Metals Company lost a potential $500 million in funding, and its
share price sank 11 percent.[110]

Impoverished nations are tempted by the deep pockets of mining corporations, and so, too, are ocean scientists. When the subject of deep-
sea mining arose in interviews, many ocean researchers and mappers
expressed a sort of helpless ambivalence about it. Sooner or later, they
expected the industry to transform the seafloor and their work along
with it. For many, it already has. Lisa Levin has estimated that roughly
half the researchers in deep-sea biology work with miners to collect
baseline data through either mining corporations or government-
funded groups. The Metals Company claims to have spent upward of
$75 million on research in the CCZ alone.[111]

   Of course, science and industry have a long-intertwined relationship. Benjamin Franklin published the first chart of the Gulf Stream,
drawing on the knowledge of whalers who tracked the current closely
in their search for blubber.[112] Bell Laboratories, the research arm of the
American Telephone and Telegraph Company, funded the first map
Marie Tharp drew of the Mid-Atlantic Ridge, not because Bell wanted
to help uncover plate tectonics but because, if you'll recall, it wanted to
know if its telegraph cables at the bottom of the Atlantic would snap.
"Even that big seminal work that we like to romanticize and think was
just this big academic endeavor, it actually had ties to industry," said

Vicki Ferrini, the head of the Atlantic and Indian Oceans Regional Center at Seabed 2030 and a die-hard Marie Tharp fan.

Richard Jenkins, the founder of Saildrone, said much the same. There simply isn't enough government money to pay for mapping the world's oceans; industry would have to play a role as well. When I asked one expert with the UN Commission on the Limits of the Continental Shelf (CLCS) whether people would ever care about the seafloor without an extractive purpose in mind, the answer was a swift and immediate no. In the 1980s and '90s, when metals prices tanked and mining corporations shelved their seabed-mining projects, funding for deep-sea research nose-dived as well. Since 2010, interest in seabed mining has started to tick up again, and so, too, has funding for deep-sea science.[113] Researchers and miners are locked in an uneasy partnership. Studying the deep sea is astronomically expensive, so experts need the funding and miners need the experts under the ISA rules that require companies to conduct baseline research as part of their exploration contracts. The ethical problems arise when miners start to dictate the questions the researchers can ask, influence the results, or block access to data.[114]

After lunch, Patricia Esquete and I wandered back to the ISA to see whether the negotiations had opened to observers again. Along the way, we passed banner after banner advertising the authority's commitment to science, diversity, and the environment. For nearly twenty years, all the baseline data that the ISA required contractors to collect was stashed away, accessible only to prospectors, the scientists behind the research, and a few ISA staff. (Occasionally some information was published in peer-reviewed studies.) That was supposed to change in 2019, after the ISA finally launched a public database. However, researchers found the data collected by miners inconsistent or incomplete. For instance, there are no resource data, which are considered proprietary, even though the nodules form a major part of the habitat. An estimated 50 percent of larger animals in the eastern CCZ live on the nodules, and so in those areas researchers are seeing only about half the picture. "While there is some very good data, there is variable quality among the mining contractors, there's no doubt about that. The CCZ is grossly under sampled overall," Craig

Smith, a deep-sea scientist at the University of Hawai'i, told *National Geographic* in 2019.[115]

The ISA also runs a training program that places young scientists from the developing world on mining ships. However, when I reached out to an African scientist who had participated in one of those training cruises, he couldn't speak about any of the science he had conducted on board due to a nondisclosure agreement he had signed with the Metals Company. NDAs are common in industry circles but foreign to scientists, who are trained to publish what they learn. Jeffrey Drazen, a professor of oceanography at the University of Hawai'i, has called the NDAs "antithetical to science."[116] The Metals Company has funded Drazen's research on mining impacts to the midwater column, and two anonymous sources told the *Wall Street Journal* that the Metals Company had warned Drazen that he might lose his funding if he continued to criticize deep-sea mining. Drazen declined to comment for the article.[117]

In 2017, a definitive rupture split the deep-sea scientific community and the ISA. For nearly two decades, scientists had studied the Lost City, a bubbling metropolis of hydrothermal vent chimneys that soar nearly 200 feet from the Atlantic seafloor. Discovered in 2000, the Lost City is covered in dense and mysterious microbial communities with roughly thirty thousand years' worth of hydrothermal activity. The Lost City might be a good candidate for solving the secret to how life first appeared on Earth. Scientists believe it could hold clues about alien life on other planets, too. In 2017, the ISA decided to lease part of the Lost City to Poland,[118] a major blow to members of the scientific community, some of whom had spent their entire careers studying the vent field and felt they had not been properly consulted beforehand. Almost seven hundred scientists have now signed a petition calling for a moratorium[119] on all deep-sea mining until more is known about the seabed—a moratorium that ISA secretary-general Michael Lodge has perplexingly called "anti-science and anti-knowledge."[120] Scientists named the Lost City after Atlantis and the myth of an advanced society destroyed by its own technology[121]—a prophetic name for a place that deep-sea mining might irrevocably damage before scientists can uncover its mysteries.

5.

On a gray December day in Rotterdam, twenty Greenpeace activists motored across the Dutch port and docked next to the towering side of a 228-meter-long ship. Appropriately named *Hidden Gem*, the former oil-drilling ship was undergoing renovations to become a subsea mining ship operated by the Metals Company and its Swiss partner and shareholder, Allseas. In order to meet the miners where they are, Greenpeace has taken to protesting ships whenever they come into port and even tailing them out to sea.[122] As the activists rappelled up the hull, the sound of hammering and drilling echoed up from inside the vast ship. Once on board, the activists unfurled a banner reading STOP DEEP SEA MINING!

That happened on the last day of the ISA Council meeting, and finally it became clear why Greenpeace's Arlo Hemphill had been so on edge all week. "Now I'm completely calm about it, but I haven't been all week. I've been kind of stressing out," he said, careful to highlight that he had had zero involvement with the action. Some of that stress was caused by well-intentioned colleagues who were worried about his safety in Jamaica. Another concern was the Metals Company, whose representative sat across the council chamber from Hemphill. If there were a negative backlash against Greenpeace, Hemphill would be its focus. Once inside the tightly controlled perimeter of the ISA compound, however, those worries seemed overblown.

What did the delegates think of the Greenpeace protest? When I asked, most hadn't heard about it and their reactions were mostly amused. "Greenpeace is always scaling things," one European delegate said with a shrug. But others did appreciate Greenpeace's laserlike ability to spotlight important niche issues. The delegates here had spent years toiling away at the ISA, working toward an elusive compromise on an issue that most people had never heard of.

Many experts and observers I interviewed declined to speculate about whether commercial deep-sea mining would become a reality in 2023. "As a researcher, I refuse to answer that question," said Aletta Mondré, the political scientist at the University of Kiel, with a laugh. "Because chances are that you're going to be wrong." In her research and writing about the ISA, she has found papers dating back to the 1950s predicting the start of the deep-sea mining industry in the next ten years. "When I got into the topic, I did what a lot of people did. [I said], 'It's not here yet, but it's going to be here *definitely* in the next three years.' And that was ten years ago."

Predicting future metal demand is always tricky, but some reputable sources do predict a time, not too far from now, when metal supplies will be low and recycling and technology will have yet to catch up with demand.[123] (Opponents say that these estimates are flawed, based on status quo consumption trends that do not factor in recycling or the evolution of alternative technologies.)[124] But even a momentary demand for more metals in service of a greater goal, such as reversing climate change, might be enough to tip the world into accepting a new extractive industry at the bottom of the ocean.

There was one person I spoke to who was happy to offer his predictions on deep-sea mining. When I told Victor Vescovo I would be attending the upcoming ISA negotiations in Jamaica, he scoffed. I would probably hear a lot of hooey, he said, but he did not use the word *hooey*. In his opinion, seabed mining would never happen, purely on an economic basis. "As anyone who has any experience in commodity metals knows—as I do, I actually invested in a copper recycling company for a while—if you start throwing a lot more supply on the market, the price goes down. So I mean, have [deep-sea miners] factored that in? I just don't know, economically, how it works."

Victor also knew firsthand just how expensive working offshore is. Stuff breaks down, stuff gets lost, like his $300,000 submersible arm that had fallen off in the Atlantic and nearly killed the Five Deeps Expedition before it began. In 2021, Greenpeace tailed a ship chartered by the Belgian company Global Sea Mineral Resources and witnessed the operation lose a bright green 25-ton nodule collector on the seafloor for several days.[125] In early 2023, scientists aboard *Hidden Gem* leaked footage of the Metals Company trialing its deep-sea mining technology in the Pacific. The video appeared to show mining waste flowing directly into the ocean, rather than being pumped down a discharge pipe and released deeper down as originally envisioned. The Metals Company called the incident a "minor overflow" that was later corrected. Not long afterward, the Metals Company's stock sank 12 percent.[126] "There's just a high chance of mechanical failure," said Victor, and "there's too many people behind seabed mining who don't know enough about working in that deep-sea environment, how hard it is."

As a businessman who has also dived to the bottom of the ocean, Victor Vescovo has a rare insight into the world of deep-sea extraction. Does seeing the seafloor with your own eyes impart a new sort of knowledge, something you can't get by looking at a map of the place? I had always hoped I might get the chance to visit the deep sea in a submersible. I had reached out to a half dozen organizations around the world to try to make it happen. Not long after I returned from Jamaica, it looked like I might finally get my chance.

CHAPTER 11

TO THE BOTTOM AND BEYOND

It is not down on any map; true places never are.

—Herman Melville, *Moby-Dick*, 1851

1.

The Challenger Deep is the celebrity deep, the deepest of the deeps, the oldest of the deeps, and it's located in the Pacific, the largest ocean on the planet. On April 28, 2019, Victor Vescovo started his descent to the Challenger Deep. If he had been free-falling through air, his trip to the lowest point in the Earth's crust would have taken approximately four minutes. In water, it took him four and a half hours. He sank gently through the pitch-black water, which grew colder and denser with every passing fathom.

The Five Deeps Expedition had gone through a lot since it had first set out on the Atlantic Ocean five months earlier. After the disastrous trip through the Southern Ocean, chief scientist Alan Jamieson had nearly quit. The downtime at home in England had led him to reconsider. "When I got home after, [I thought,] 'I'm going to quit this.' But then I thought, 'All my gear is still on the ship. The only way to get it off is I'm going to have to fly to Indonesia and then pack in front of all the guys,'" he said wistfully. "Half the crew are the same age as me and Scottish. It's bizarre. It had the feel of you and your mates have just stolen a ship and taken it around the world." He eventually decided to return to Five Deeps for the leg through the Indian Ocean, where Victor would dive the third deep, the Java Trench. Afterward, Jamieson was rewarded with his first dive to the Hadal Zone—something that brightened his mood considerably.

"Java was amazing," he remembered. All but ignored by scientific expeditions until the 1950s, the Indian Ocean is the least scientifically well understood of the five oceans, and Jamieson was poised to make some big finds.[1] The scientific landers captured new species, filmed new behavior in already discovered species, and videoed the deepest octopus ever recorded on film, extending the octopus's range nearly 6,000 feet deeper and to 99 percent of the global ocean.[2] Jamieson also spotted a new species with his own eyes for the first time: a tunicate sea squirt that he nicknamed Snoop Dog for its doglike appearance.[3] "It's really, really quite brilliant. We were only [in Java] for five days, and we learned more about the Indian Ocean than we have in the last fifteen years." Jamieson's concerns about the scientific credibility of Five Deeps seemed to ebb after the Indian Ocean.

Cassie Bongiovanni and her mapping team made their own discoveries in the Indian Ocean as well. There were two candidates for the deepest deep: the Java Trench and the Diamantina Fracture Zone. (A fracture zone is a rupture in a single tectonic plate, while a trench is a meeting point between two plates.) Neither deep had stellar maps, with much of the existing data drawn from satellite predictions.[4] Together, a team including Aileen Bohan, two other GEBCO mappers, and a ship officer on board, mapped the trenches, deployed the landers, and discovered that the Java Trench beat the Diamantina by a narrow 223 feet.[5]

At last the ship pulled into the Pacific Ocean, where its crew would find out whether *Limiting Factor* could withstand the deepest deep in the world. Every berth on *Pressure Drop* was full to watch Victor make history. Just three men had reached the bottom of Challenger Deep before: the two-man team in 1960 and James Cameron in 2012. Neither team had had an ocean mapper on board or such a sophisticated sonar with them. For the dive in 1960, the crew had tossed blocks of TNT over the side and timed how long it took the explosion to echo back to the ship's hydrophone. "We did not care about exact depth measurement, only that 14 seconds was deeper than 12 seconds," Don Walsh, the US Navy lieutenant on board, wrote later in *Scientific American*. Together, Walsh and the Swiss engineer Jacques Piccard reached an ultimate depth of 10,916 meters (35,814 feet). For the next fifty years, that was the record to beat. When James Cameron dived Challenger in 2012, he reached a slightly shallower depth of 10,898 meters (35,756 feet) but won the consolation prize of deepest solo dive in the world.

The Mariana Trench is one of the best-mapped trenches in the world. In 2017, Cassie's alma mater, the Center for Coastal and Ocean Mapping at the University of New Hampshire, mapped the Mariana Trench with the EM 122, a precursor of the EM 124.[6] Those maps eventually became part of the USGS Law of the Sea project, carried out with the express aim of extending US territory offshore. These maps are some of the most rigorous and heavily scrutinized in the hydrographic world.

Cassie Bongiovanni didn't look at any of them. "With Challenger [Deep], I didn't want to have any preexisting information. I wanted to treat it like every other trench that has never been mapped before. If I knew where everybody else said the deepest points were, you could subconsciously bias yourself into saying that's what the deepest point is," she told me. According to Cassie, James Cameron had touched down on what appeared to be a bridge or a hill on the Challenger Deep. Cassie, meanwhile, went on to discover a deeper spot just two nautical miles away; if her calculations were correct, she could direct Victor to the deepest point on the planet and earn him two world records in a single day: deepest dive *and* deepest solo dive in history.

2.

One of only three men in the world to see the Challenger Deep with his own eyes had taken a berth on board *Pressure Drop*. In 1960, when navy lieutenant Don Walsh had dived the Challenger Deep for the first time, the stakes had been considerably higher, both for his personal safety and for global politics. Those were the early days of the Cold War, and the Soviet Union had pulled ahead in the Space Race, launching the first satellite and the first man into space. The United States, keen to prove its technical prowess, purchased an underwater craft called a bathyscaphe, created by the Swiss physicist Auguste Piccard. In the 1930s, Piccard had built helium balloons that had enabled scientists to reach new heights and take stratosphere-busting measurements. Thirty years later, Piccard would send men in the opposite direction: to the bottom of the ocean.

The bathyscaphe consisted of a pressurized steel sphere with large tanks mounted on top that held more than 30,000 gallons of aviation fuel. The highly flammable fuel just above his head was a slight concern, Walsh remembered.[7] Lighter than water, the fuel would not compress under high pressure and turn the bathyscaphe buoyant. To start the descent, another set of tanks at the top of the bathyscaphe took on water, sinking the submersible into the sea. Once the mission had been completed on the ocean floor, the bathyscaphe would jettison 8 tons of steel shot and rise up to the surface. One spot aboard the bathyscaphe went to Auguste Piccard's engineer son, Jacques; the other to Don Walsh. In case the mission backfired and embarrassed the Americans, not to mention killed the two men on board, the US Navy kept the whole thing hush-hush, according to Walsh.

In January 1960, the USS *Lewis* quietly sailed out of Guam and headed for the Challenger Deep, about 60 miles away. The USS *Wandank* followed, towing the very delicate bathyscaphe at a crawling 5 knots.[8] *Lewis* arrived first and began to search for a deep spot in the Challenger Deep, tossing TNT over the side and timing how long it took for its blast to echo back. By the time *Wandank* arrived, bathyscaphe in tow, the team had found a promising site. Between breaks in the 25-foot waves, Walsh and Piccard jumped from *Lewis* into the bathyscaphe. Down, down, down they descended, until, hours later, at 9,448 meters (31,000 feet) deep, they heard a muffled bang. Startled, the two men looked around, trying to find the source of the noise, but everything seemed to be functioning fine. They continued to dive, subjecting the bathyscaphe to greater and greater pressure. Finally, five hours after they had begun, they glimpsed a muddy seafloor loom up below.[9] At their ultimate depth, 16,000 pounds pressed onto each square inch of the bathyscaphe, similar to the atmospheric pressure found on Venus. Walsh flicked on a viewing light to look out the porthole of the exit hatch and saw a huge crack across the acrylic window. There was the cause of the bang. Luckily, Walsh and Piccard realized that the window was so tightly sealed by the water pressure that it posed no danger.

"After we landed Jacques and I shook hands and expressed our feelings of relief and joy," Walsh later recounted. "Our small Project Nekton team had said we would do it and we did!"[10] They stayed just twenty minutes at full ocean depth before starting their ascent. From there, both men were whisked off to Washington for a hero's welcome and a meeting with President Dwight D. Eisenhower. Don Walsh expected that he and Piccard would become the first in a long line of deep-diving aquanauts.

Just two short years later, in 1962, the new president of the United States, John F. Kennedy, gave a rousing speech on the next frontier the United States should conquer. "We choose to go to the moon," he boomed to a crowd gathered in a Texas stadium. "We choose to go to the moon in this decade . . . because that goal will serve to organize and measure the best of our energies and skills, because that challenge is one that we're willing to accept, one we are unwilling to postpone. . . . And, therefore, as we set sail we ask God's blessing on the most hazardous and dangerous and greatest adventure on which man has ever embarked." For a brief moment, the dive in the Mariana Trench had compelled people to contemplate the depths below. Kennedy's speech harnessed the public's imagination and aimed it skyward. Outer space swiftly replaced the ocean as the next exciting frontier.

That worried the author and ocean lover John Steinbeck, who predicted that exploring outer space would distract from the work and wonder of documenting our own planet. "It does seem unrealistic, unreasonable, romantic, and very human that we indulge in these passionate pyrotechnics when, under the seas, three-fifths of our own world and over three-fifths of our world's treasure is unknown, undiscovered, and unclaimed."[11] Throughout the latter half of the twentieth century, those concerns came true. Public funding for ocean exploration dwindled, while the public's passion for exploring outer space seemed to know no limits.

Today this rabid thirst for everything outer space is easy to track online, where space travel is searched four times as often as ocean exploration.[12] The social media accounts of NASA, Blue Origin, and SpaceX count tens of millions of followers. Similar accounts for ocean exploration are lucky to reach a million. In the lead-up to SpaceX's first launch in 2020, millions of viewers tuned into a live feed of the launch pad in South Texas. "Even in the middle of the night, when there's absolutely nothing going on, and you're staring at a launch stand with nothing on it, there will still be two thousand people in there watching, and a few dozen people chatting about whatever," the owner of the YouTube channel told a Texas magazine in 2020.[13]

If it were only a popularity contest between space and the ocean, the numbers wouldn't matter so much. But of course popularity translates into public funding. The gulf of interest between exploring ocean and outer space helps explain the steep difference between NOAA's small budget and NASA's generous one.[14] Today, most ocean researchers have stopped trying to fight the tide and learned to embrace the public's love of space travel. Ocean research vessels are named after famous astronauts, such as R/V *Neil Armstrong* and R/V *Sally Ride* at the Scripps Institution of Oceanography. Marine researchers drop buzzwords such as *ocean worlds* into grant applications, boosting the value of ocean life by tying it to the search for extraterrestrial life in outer space.

3.

The ocean and outer space are the two ideological endpoints in exploration, so naturally they share some of the same goals and technology. Deep-sea submersibles are built to withstand cold temperatures and high pressures, as well as the corrosive efforts of salt water. Once on their journeys, they perform routine tasks again and again, with little or no opportunity for repairs. Extraterrestrial rovers and spacecraft endure similarly extreme conditions and are dispatched on lifelong journeys to the ends of the universe, never to be seen by humans again.[15] Astronauts and aquanauts train for their epic adventures in similar ways, too. At NASA's Neutral Buoyancy Laboratory in Houston, astronauts practice spacewalking in a mock-up of the International Space Station submerged at the bottom of a mammoth pool.

Only the elite can afford to pay for such activities, so the fields share another overlap in *who* gets to do the exploring. Until recently, that meant the highest achievers in the military and scientific worlds. But after decades of neoliberal capitalism, hobbled government agencies,[16] and the wealthiest people and corporations paying little to no tax, the richest people in the world can now launch private exploration companies that rival the US government's. For now, the clients for deep-sea and space travel are other ultrarich patrons or government agencies, such as NASA contracting SpaceX to carry astronauts to the International Space Station. But the spacefaring entrepreneur Elon Musk—at one point the richest man in the world, whose car company, Tesla, paid zero federal income tax in 2021[17]—dreams of a day when the masses can afford a trip to Mars. Of course, everyone would be better served by less flashy ambitions, such as a sustainable future here on planet Earth.

The two fields also share the hunt for extreme life that thrives in severe conditions found in the deep sea and on faraway planets. At NASA's Jet Propulsion Laboratory in Pasadena, California, there's a whole department devoted to exploring the overlaps between life at the bottom of the sea and on other "ocean worlds": moons with water, such as Jupiter's Europa and Saturn's Enceladus. Marine scientists have learned to hitch their research to outer space to boost funding, and NASA has done the same, expanding and promoting its earth science research as the environmental movement has grown and pushed for public money to be put to use on this planet.[18]

The cruise on *Nautilus* that I took up the California coast was a good example of that. NASA had funded the expedition thanks to a series of deep dives looking for extreme life-forms. Immediately off the coast of Santa Barbara, the seafloor plunges 500 meters (1,640 feet) to a flat basin bottom before it rises again to the Channel Islands. This bathtub shape creates a stagnant deep-sea environment that is perfect for sampling and studying life-forms that can survive under high pressure and with little to no oxygen or sunlight. Those life-forms just might look like the alien life we could find on distant planets. Or they might serve as good test subjects to send into space on a long galactic journey.

In the end, I never found my ride to the ocean floor. Watching ROV *Hercules* scour the Santa Barbara Basin was the closest I came to exploring the seafloor myself. Patrick Lahey, the ever-accommodating CEO of Triton Submarines, tried to take me on a test dive in the Bahamas that never materialized. Other options bubbled up and then faded away. I always knew my chances were slim. Far more worthy scientists had waited decades for a seat aboard a submersible. There were many paying customers ahead of me, too. I asked Victor Vescovo for a ride in *Limiting Factor*, but he was charging customers $750,000 for a dive— way outside my price range but a steal compared to the $28 million Blue Origin charged its first client for a ride to the edge of space.

The ROV dives I watched off Santa Barbara gave me a glimpse into the commonalities between ocean and space exploration. Scientists have spent years studying and sampling the basin, making it some of the most studied seafloor in the world, yet they're still learning new and novel things about deep-sea life all the time. During the days I spent on board *Nautilus*, I watched scientists pull up deep-sea clams with a certain bacterium in their gills that allowed them to survive for up to ten months at a time without oxygen. They were also on the hunt for foraminifera (known more colloquially as forams), a diverse group of microscopic shell-forming organisms that look quite pretty under an electron microscope, some shaped like pine cones, others like twisted tulip bulbs. Forams likely evolved more than 541 million years ago during the Precambrian period, so these single-celled organisms provide a glimpse into early life on Earth. They might also show what the future holds for the ocean. Forams flourish in so-called dead zones, which are spreading at sea due to climate change and pollution. Most people associate dead zones with river outlets, where runoff from fertilizer and human waste leads to phytoplankton blooms that strip oxygen from the water. However, dead zones also occur naturally at the bottom of stagnant ponds and bodies of water, such as the Santa Barbara Basin. The term *dead zone* is a bit of a misnomer, because life persists there, as the forams prove. As dead zones spread in the deoxygenating ocean, there might be many more forams at sea.

For the first dive of the cruise, the scientists were searching for bacterial mats where forams and other fascinating microbial communities gather. There were so many excited people on board who wanted to watch the first dive that a half dozen of us crowded into an overflow viewing room below the control van where the pilots directed ROV *Hercules*. The overflow room was actually *Nautilus*'s TV lounge, decorated in a mix of nautical and NASA. One side of the room was all portholes, wood paneling, and leather couches, the other a bank of computer screens filled with streaming data and live video feeds. As the dive unfolded, the researchers watched spellbound. It began, of course, at the surface of the sparkling blue sunlit waters. Within seconds of the ROV's sinking, all light from above disappeared and the light from its headlamps grew brighter and almost eerie in the darkness, illuminating millions of particles swirling past. Those particles could mean life or death in the sea. They could be plankton; they could be microplastic; they could be marine snow—such a poetic term for floating feces or dead microscopic animals.

Watching the particles scroll past the black ocean background began to remind me of something. Victor Vescovo had described seeing a similar view while descending into the deep ocean, but now I was watching it myself in real time. Then it came to me what it looked like: the night sky! We were sinking into the ocean, yet it looked as though we were roaming in outer space, each particle a star. "It's like jumping into light speed," one scientist behind me said, right on cue.

"Mat or no mat?" asked Nicole Raineault, the expedition leader. That was the question. The Santa Barbara Basin flushes itself sporadically, washing away the life that builds up on the bacterial mats. Without a mat, there would be no microscopic life for the scientists to sample and study. Raineault had taken a seat on the carpet, closest to the TV screen, and she worked her way around the room, taking guesses from all the researchers.

Watching the seafloor loom up underneath the ROV felt a bit like watching the sun rise. Staring at the dark night sky, you barely notice it lightening until suddenly brightness is shining all around you. The same thing happened in the deep sea: the black ocean lightened ever so gradually from black to blue, and then I realized that *Hercules*'s headlamp beams were bouncing off the bottom. Then everything came into focus very quickly; the seafloor sharpened, the texture resolved, the researchers leaned forward expectantly. The moment of truth had arrived.

One researcher, who had been too cautious to guess earlier, was perched right in front of the TV, his nose almost touching the screen. *Hercules* paused a moment, hovering 65 feet above the murky bottom. The researcher waved his hand impatiently—"C'mon, already! Get going!"—and then *Hercules* started to sink again. A flat, smooth, grayish bottom pulled into view, speckled with bright white patches. "Oh, yeah!" the researcher said excitedly, pulling back from the screen and grabbing his phone. "Those are mats." The room broke into cheers and claps. It was the seafloor and the extreme life we'd come all that way to see.

4.

For decades, marine conservationists and scientists have racked their brains trying to figure out why the public prefers to explore outer space over the ocean, despite the many similarities between the two. The reasons seem to be political, psychological, and "very human," as John Steinbeck pointed out all those years ago. The Cold War certainly played a role in funding space travel. (Of course, a sizable amount of military funding spilled into deep-sea exploration, too, especially during Marie Tharp and Bruce Heezen's time.) In war, the tactical advantages of patrolling the skies and seas are clear. The United States and the Soviet Union wanted to spy and surveil, to launch missiles and warheads, but they also had to inspire the public with something more than militaristic ambition. The moon shot was perfect. "The same technology that transports a man to the Moon can carry a nuclear warhead halfway around the Earth," Carl Sagan wrote in a famous 1989 essay supporting nuclear disarmament.[19]

But NASA's moon shot also tapped into a deeper reality of why people prefer the skies over the seas. Fear plays a role in what the public chooses to support, as wolf conservationists know all too well.[20] In nearly every seafaring culture, humanity's fear of the deep is personified as a sea monster, whether it's the giant Kraken in Norse mythology or the Yacumama serpent slithering through the Peruvian Amazon.[21] No matter how many statistics you cite on the improbability of a shark attack, sharks still play the role of bloodthirsty hunters of humans in the popular imagination today. A sizable part of society will always avoid deep water out of an instinctual fear—but fear of what, exactly? Victor might have glimpsed the answer during his expedition to the deepest points on the planet. Not a single particle of light penetrates the ocean bottom, and Victor spoke of seeing the absolute blackest black he had ever seen outside his view port. "It was wild," he wrote me. "Absolutely zero depth perception. Total and utter void." Sometimes he would turn off all the lights inside the submersible, cup his hands to the glass, and stare out into the abyss. The moment reminded him of the famous "staring into the abyss" quote from the German philosopher Friedrich Nietzsche: "He who fights with monsters should be careful lest he thereby become a monster. And if thou gaze long into an abyss, the abyss will also gaze into thee." Perhaps here lies the real reason we fear exploring the deep sea: its darkness might destroy us.

Religious mythology can't be discounted, either. Heaven is up, just like the sky; hell is down, just like the ocean. Believers pray to a god above and look to the heavens for guidance and protection. They dread what lies beneath, where they'll come to rest in an earthen grave or, worse, pay for their sins in hell. Regardless of whether you believe in a God or the afterlife, the religious orientation extends to the nonbeliever as well. Looking up at the stars and imagining a better place beyond the horizon is a uniquely human trait. It's the same urge that compels us to travel to new lands and wonder about lives and perspectives beyond our own. But it also has the unfortunate downside of driving us to distraction, to hunger to be the "first" and go astray from the difficult work of looking deeper inside ourselves and asking what a better world might look like right here, right now.

Holed up in the solitude of his cabin on Walden Pond, the American writer Henry David Thoreau ruminated on humanity's endless wandering. He lived through what's been called the second great age of discovery during the nineteenth century, when explorers and expeditions were setting out for the Arctic, the Amazon, Antarctica, and many other remote, uncharted landscapes. He was fascinated by those journeys and kept careful track of the discoveries, but he also seemed repelled by the insatiable urge to gobble up new terrain. "It is easier to sail many thousand miles through cold and storm and cannibals, in a government ship, with a hundred men and boys to assist one, than it is to explore the private sea, the Atlantic and Pacific Ocean of one's being alone," he wrote.[22] Inner soul-searching must accompany outward exploration, or else we fall into a trap of ticking off new frontiers like an endless shopping list. Thoreau's words also anticipate a time when we will have moved past exploring the ocean—one of the most cutting-edge pursuits of his day—and on to some new, more exciting frontier.

This is the situation we find ourselves in today. Around a quarter of the ocean floor has been mapped, less than 1 percent of the deep sea has been explored, but billions of dollars are being poured into space exploration and, even more recently, space militarization. The ocean, its inhabitants, geography, and meaning, are only barely fathomed, yet the sea stirs up only a fraction of the excitement as the other frontier does. The stiff headwinds of money and geopolitics have played some part, but so too have humanity's psychological blind spots. By the early twenty-first century, no human had ever dived the deepest points of all five oceans. We didn't even know where those points were until Victor Vescovo and Cassie Bongiovanni set out to find them.

5.

When *Limiting Factor* surfaced from the Challenger Deep, Victor Vescovo opened the hatch and thrust out a hand with four fingers raised:

one for each deep conquered. Just one more to go. In total, he had spent three hours at the bottom of Pacific Ocean. The crew had dropped scientific landers in strategic spots and Victor had roamed around to them, covering all the potential deepest points on the seafloor so that no one could doubt his claim later on. Baited with food, the landers drew in gelatinous creatures, and Victor watched as deep-sea sea cucumbers, known as holothurians, and sea worms slowly assembled to nibble at the food in the traps.[23]

The depth gauges aboard *Limiting Factor* recorded his deepest depth as 10,928 meters (35,853 feet), besting both James Cameron's 2012 dive and the inaugural Walsh-Piccard dive in 1960. A press release went out announcing Victor's two new world records—news that naturally made headlines around the world. It also ruffled the feathers of James Cameron, who was filming the next installment of *Avatar* in New Zealand. Behind the scenes, Cassie had her doubts about the depth reading aboard *Limiting Factor*. Victor had pushed for the higher, more impressive number from the depth gauge, while Cassie wanted a lower, more conservative number based on her EM 124 readings. "'Victor, if you say 10,928 [meters], I don't know if I can support that,'" Victor remembered Cassie telling him. Eventually he agreed, and in a later press release, Five Deeps rounded the number down to the statistically safer 10,924 meters (35,840 feet) with a 15.24-meter (50-feet) margin of error in both directions.[24] But by that time, the damage was done. The film director had contacted major news outlets, requesting to speak and question Victor's record. "You can't go deeper," he told the *New York Times*. "It's flat and featureless. So his gauge may read differently from mine, but he can't say he's gone deeper."[25]

Larry Mayer, the director of the Center for Coastal and Ocean Mapping at the University of New Hampshire, assured Victor that Cassie had indeed surveyed the Mariana Trench correctly. James Cameron might have seen flat seafloor out his submersible's porthole, he explained, but he could barely see 200 feet away.[26] Just beyond that line of sight, the seafloor could slant either up or down. Though Victor seemed to relish the media attention James Cameron had brought to the dive, the spotlight was extremely distressing for Cassie. On her first job out of university, one of the world's most famous directors had publicly dismissed her work in the *New York Times*, complaining that "something can become part of the public record without substantiation."

Over the coming years, *Limiting Factor* would go on multiple dives in the Challenger Deep. Victor Vescovo employed a trained ocean mapper and relied on the world's most advanced sonar system, along with scientific landers, to verify the depths. James Cameron hadn't done those things. But I found myself most convinced by the ocean mapper, checking and double-checking and triple-checking her work. "I'm not in it for personal gain," Cassie told me earnestly. "I'm going to tell you what the data says. And that's it."

The squabble highlighted yet another reason why the public seems to gravitate toward outer space. A space mission may be freighted with nationalistic goals, but success in outer space somehow transcends earthly politics. No matter a person's nationality, he or she can gaze up at the sky and wonder "What if?" Even when diplomatic relations between countries break down, astronauts of all nations continue to collaborate on the International Space Station.* The ocean, meanwhile, does not inspire such a generous fellow feeling. The depths are steeped in factionalism and politicking as they became a graveyard for human warfare.[27] At the International Seabed Authority, nations have spent years arguing over regulating an industry that might not even be profitable but will most certainly damage an ecosystem that belongs to all of humanity. To an outside observer, the dispute between James Cameron and Victor Vescovo looked like another bad-news story in the ocean world: two rich white men bickering over a world record that most people had never heard of.

---

* In 2022, Russia announced that it would quit the International Space Station after 2024—a sign that low Earth orbit may have exited the early collaborative period and entered the next competitive phase in exploration: https://www.cnn.com/2022/07/26/world/russia-quit-iss-scn/index.html.

6.

Pressure Drop sailed on, moving into the Arctic Ocean for the fifth and final dive of the expedition. In the summer of 2019, the crew watched from the chilly decks of the ship as Victor descended to the bottom of the Molloy Hole, about 170 miles west of Svalbard, Norway.[28] He spent two hours sinking to the bottom, two hours cruising around the Arctic seafloor, and another two hours to resurface. When he cracked open the submersible's hatch, the sound of cheers and popping fireworks greeted him. He had just become the first person to dive to the bottom of the Arctic Ocean as well as the first to dive the deepest points of all five oceans. A wave of relief washed over him, the same kind he remembered feeling at the top of Mount Everest. "Any number of things can derail these very difficult expeditions. So when you actually finish it, there's an immense amount of relief," he said. That relief quickly turned to excitement: he had done it—something that no one had ever done before. And then, just like that, the Five Deeps Expedition was over. After two more science dives, the ship scurried back to the safety of a Scandinavian fjord right before a storm swept in. Down in the dry lab, Victor added the final deep to a wall tracking all the dives: "Molloy Hole—5,550 meters."[29]

From there, the ship kept sailing to London, England, where the Royal Geographical Society invited the team to speak about the record-breaking expedition. Inside the society's historic ballroom, Alan Jamieson stood on a stage and rattled off a series of numbers: forty thousand biological samples, close to 5 million feet of water data collected, more than five hundred hours of deep-sea video footage, and a biodiversity survey of the Abyssal and Hadal Zones. It would take years for his team to parse the data collected and understand what it all meant. One day, he predicted, the data would fill in the missing gaps about the evolution of life on Earth.

Cassie spoke, too, about the 265,700 square miles she had mapped, much of it for the first time in history; the hundred new seafloor features she had discovered; and the new surveys of important trenches, fractures, and subduction zones she had conducted around the world.[30] It was a career highlight, but as soon as she stepped down from the stage, she began to worry about her next steps. "Everybody was coming up to me saying, 'You've peaked! Where can you possibly go from here?'" she remembered. Over the course of a year and a half, the now twenty-seven-year-old mapper had gone on a mind-bending journey around the world; suddenly the emotional strain of it all came crashing down on her. "I left the ship, and I was on my own for three days, for the first time in a really long time, and I just started crying," she said. She was unsure whether she would see the ship or the crew again as the expedition disbanded and everyone went his or her separate way.

Victor had always planned to sell the operation after he finished diving the Five Deeps. The Triton Hadal Exploration System, including the submersible, the ship, the EM 124 sonar, the support vessels, and the landers, was put onto the market at $51 million. The Triton team had a financial stake in the sale and reached deep into its network to find a buyer. They contacted nearly three hundred ocean research organizations around the world and thousands of single-family investment offices and mailed glossy brochures to captains, owners, and architects across the superyacht industry. Various high-profile names were floated: Ray Dalio, the billionaire hedge fund manager turned ocean conservationist behind OceanX; the Australian billionaire and philanthropist Andrew Forrest; a rear admiral from NOAA. No one came through. *Limiting Factor* had achieved everything it had set out to do. It was a one-of-a-kind submersible that could dive any seafloor anywhere in the world and do it repeatedly to full ocean depth. That was still not enough of a draw for people with all the money in the world to buy it.

There were other factors that might have played into the tepid response, including the dawn of the COVID-19 pandemic and travel restrictions. Still, the utter lack of interest in the world's deepest-diving submersible surprised the Five Deeps team. "I really can't explain why there was not as much interest in acquiring the system, either [by] government or high-net-worth individuals," Victor said. "Over the last couple of years, the whole team, including our scientific groups that have been on the ship, it's a constant lament: 'Space is exciting and new and loud.' . . . But the ocean just doesn't get the funding." After the surge of investment in ocean initiatives during the late aughts and early 2010s, the enthusiasm for ocean exploration had plateaued.

With no serious offers for the Triton Hadal Exploration System, Victor embarked on a grab bag of expeditions, from searching for a French submarine lost off the Mediterranean coast in 1968 to reaching the deepest points of the Mediterranean and Red Seas. He returned to the Challenger Deep, where he took the first woman and astronaut, the first US marine, and the first person of Asian descent to the deepest point on the planet. He made the first-ever ascent of Hawaii's Mauna Kea, the world's tallest mountain from seafloor to summit, with the surfing scientist and Hawaiian native Clifford Kapono. Together, the pair embarked on a triathlon of a world record: diving to Mauna Kea's base 4 miles beneath the surface in *Limiting Factor*; paddling an outrigger canoe 27 miles to shore; biking 37 miles up the mountain until the road ran out; and then hiking the last 6 miles to the summit.[31] All those expeditions served as a marketing tool for the Triton Hadal Exploration System; it showed the world what the submersible could do.

Back at her parents' house in Dallas, Cassie relaxed after her year and a half traveling the world and started to think about the next steps to take. In the end, she decided to join Victor Vescovo's Caladan Oceanic Expeditions and turned her attention to the next onboard mapping mission. One voyage in particular captured her heart: mapping the Pacific Ring of Fire. In August 2020, she set out on *Pressure Drop* to survey the volatile trenches rimming the world's largest ocean. Most earthquakes and volcanic eruptions emanate from the Ring of Fire, including the 2011 earthquake off the coast of Japan and the underwater volcanic eruption near Tonga in 2022. Despite all the destruction that the Ring of Fire has caused over the years, most of its temperamental trenches are uncharted, little understood, and constantly in flux. During the expedition, Cassie mapped eight trenches, including the Philippine, Yap, Palau, Kuril-Kamchatka, and Aleutian trenches, for the first time. ("Like Pokémon, I'm collecting all the trenches," she joked.) In all seriousness, though, she respected Victor's decision to map the Ring of Fire. "He doesn't have an obligation to do it. It's wildly helpful for the global collective to use his system," she said. There were still record-setting moments along the way, but the latter half of the expedition was devoted to mapping the seafloor. Cassie believed that Five Deeps had transformed Victor: he wasn't just some record-setting explorer always chasing the next horizon; he stood firmly on the side of science, too. Alan Jamieson, who had also taken part in the expedition, felt the same.

"I'm coming into this thinking, 'If this isn't scientifically brilliant, I'm out.' And [Victor is] coming into this thinking, 'I'm an explorer. I don't care about science.' At some point, we both crossed into each other's thing," said Jamieson. During the trip, Cassie passed her own milestone: 1 million square kilometers (386,102 square miles) mapped at sea by herself and her mapping assistants. All of the maps would be donated free of charge to Seabed 2030. The final tally came out to over 400,000 square miles of seafloor, an area bigger than her home state of Texas.[32] After each cruise, she retreated to her parents' home in Dallas to process the terabytes and terabytes of new ocean maps. Victor lived not too far away. One day in late 2020, he visited Cassie's parents' home to check out new mapping software she was installing. It was then that she resigned as the lead mapper for Caladan Oceanic. "I've always wanted a little bit of adventure, but I have other goals, too," she said.

The years at sea had changed her. At the University of New Hampshire, she had written her master's thesis on improving NOAA's formula for surveying the highest-priority seafloor—a perfect topic for landing a government job. "Charting is [all about] 'Hey, let's keep our mariners safe, and how do we do that efficiently with the limited resources we get?'" she explained. The time on board *Pressure Drop* had shown her that ocean mapping could contribute to so much more than safe navigation. Ocean maps uncover submerged human history and empower small communities living along uncharted coastlines; they open up new questions that we never dreamed of asking. Five Deeps was just the beginning, she realized. She wanted to keep exploring and using ocean maps to unpack the thorniest questions in biology and geology. "This is only a taste of the science that's out there," she promised.

Around the same time Cassie parted ways with Victor, a position opened at New Zealand's National Institute of Water and Atmospheric Research (NIWA). If she got the job, she would work under Kevin Mackay, the SCUFN member who also leads the Seabed 2030 South and West Pacific Ocean Regional Center. Cassie had personally surveyed 203,861 square miles in that part of the ocean, and she would get to work directly on maps she collected herself. "It sounds like the position is just taking the next step with the data," she explained. In the end, though, she settled closer to home. She now works as an engineering scientist at the University of Texas at Austin.

In the summer of 2022, the first crop of deep-sea tourists visited the *Titanic*, paying $250,000 for a submersible ride to the storied shipwreck lying 13,123 feet deep in the Atlantic Ocean.[33] Now that the deep sea is starting to host its first tourists, miners can't be far behind. Not long after Victor Vescovo broke two world records in the Challenger Deep, China sent a crewed submersible to the bottom of the Mariana Trench. The chief designer of the Chinese submersible later told newspapers that the team was building a "treasure map" of the deep sea.[34]

By late 2022, Victor had grown increasingly frustrated with navigating the byzantine rules around exploring the ocean. Apart from shipwrecks such as the *Titanic* that lie in international waters, the most visually appealing and geologically interesting seafloor is closer to shore and within the EEZ of the nearest coastal nation. A small, seemingly insignificant line in UNCLOS allows nations to regulate scientific research conducted within their EEZs—a huge territory that reaches out to 200 nautical miles (230 miles) from the shoreline. But each nation interprets "scientific research" a little differently, and Victor found that his mapping and diving requests were ignored or rejected on an ad hoc basis by the governments in charge. "I am very jaded by pursuing scientific research in countries' EEZ with the permitting required," he told me at the time. His ambitions appeared to be running headlong into a historical pattern in exploration: as more pioneers flood over a frontier, the terrain becomes crowded, exploration becomes more complicated, and the frontier is no longer frontier. Explorers then tend to seek out new, unrestrained frontiers.

In June 2022, Victor joined the fifth paid flight to space with Blue Origin,[35] the space tourism company founded by Amazon's Jeff Bezos—a ride he called "ten minutes of pure unfettered joy." A few months later, in November 2022, he sold Triton Submarines' Hadal Exploration System to the American video game billionaire Gabe Newell and his ocean exploration research organization, Inkfish. In the future, Victor planned to shift his interest in the ocean to areas that would require less permitting, such as investing in autonomous ocean technology or diving shipwrecks. In four years of exploring the ocean, *Pressure Drop* had mapped more than 1.5 million square miles of seafloor, about half the area of Brazil.

Does mapping a place equal knowing a place? On my phone, I can scroll over a Google map of a neighborhood and construct an idea of the place, but after I visit it in real life, my understanding of it grows and changes, informed by the unexpected. A map of the seafloor is the same: it's just the first step in getting to know the ocean, and in a way, it raises more questions than it answers. After that comes a never-ending cascade of mysteries to solve, debates to settle, creatures and features to name and describe, shipwrecks to investigate, human history to uncover. Perhaps we may even solve the greatest mystery of all: where we came from and how we got here. The birthplace of all life on Earth is down there somewhere, hidden in the cracks and fissures on the seafloor, where the first chemical sparks of life clicked and caught fire. Mapping the seafloor is also a quest to understand ourselves.

EPILOGUE

A map of the world that does not include Utopia is not worth even glancing at.

—Oscar Wilde, "The Soul of Man," 1891

While writing a book about the quest to map the ocean by 2030, I often stumbled upon other goals and predictions for that same year, so I started to collect them. By 2030, ocean conservationists were calling to protect 30 percent of the ocean; automated cars would drive 95 percent of the miles in the United States; the Group of Seven (G7) countries would stop using coal; global climate change would force a hundred million people into poverty; the insect protein market would become an $8 billion industry. On and on and on: these deadline-driven predictions have become something of a fad in research and policy circles. They have a way of focusing the mind, allowing us to project forward and see how far we have to rise (or fall) in the years to come. Some are hopeful visions, some are bleak prophecies, but together I started to read them as a sort of map to the future.

The scientific community has waited a long time for Seabed 2030; surely the discoveries in archaeology, biology, and geology will be mind blowing and planet altering, and the wider benefits for environmental management and safe navigation indisputable. In 2022, Seabed 2030 released the latest GEBCO grid with 23.4 percent of the seafloor mapped in high resolution, adding an area roughly the size of Europe to the map.[1] The scientists are ready, but is the rest of the world? Given humanity's historic neglect and fear of the ocean, I wondered whether our species is really the best custodian of a map of the largest

habitat on Earth. Seabed 2030 won't open the seabed to deep-sea mining, but there's nothing to stop new extractive industries taking shape
as resources on land become scarce. Uninhabited landscapes are still
"landscapes that can be charted," writes the journalist Stephen Hall,
"producing maps on which human destinies can unfold in less than benign ways."[2] At the International Seabed Authority meeting in Jamaica,
I saw one of these less-than-benign destinies unfolding: government institutions and international law buckled under the weight of corporate
entities pushing to mine seafloor that legally belongs to all humanity.
The seafloor's natural inaccessibility just might offer better protection
than any rules or regulations humans can devise.

It's helpful, then, to look at how two extreme, sparsely populated
frontiers have fared since they were conquered by cartography in the
twentieth century: Antarctica and the Amazon. One is a frozen continent whose existence the ancient Greeks mythologized long before it
was sighted in 1820; the other a hot, humid jungle where Amazonian
tribes live uncontacted and poisonous snakes slither. These two places
became the high points in exploration during the nineteenth century.
Both are wreathed in water that made them impervious to traditional
surveying, not unlike the seafloor today, shielded by miles of ocean.
Antarctica has thousands of feet of solid ice pack hiding the land's true
shape. The Amazon has low, dense clouds hovering over the jungle
that block aerial surveys. Was Antarctica a new continent? Where did
the Amazon River begin? At the Royal Geographical Society in London, debate raged over the places that resisted mapping. By the mid–
twentieth century, though, advances in surveying had unveiled both
terrains. In Antarctica, airborne radar echo sounding enabled planes
to sweep across the ice pack beaming radar that collected a continuous
profile of the terrain beneath the ice.[3] In the 1960s and '70s, side-
looking airborne radar (SLAR) used lasers to cut through the cloud
cover of the South American rain forest and create the first detailed
maps of the Amazon.[4]

The maps now made, here is where the story of the two frontiers
diverges. By the time Antarctica was properly surveyed, the world's
seventh continent was protected by an international treaty that en-
trusted its management to science until 2048. The first complete maps

of the Amazon, however, opened up a different path. Brazil's Ministry of Mines and Energy used them to lay out the rain forest's land use potential, which expedited clear-cutting for agriculture. In 2021, Brazil pledged to end and reverse deforestation in the Amazon by 2030; however, that same year, deforestation in the Brazilian Amazon reached a record fifteen-year high.[5]

The international seafloor is more akin to uninhabited Antarctica than it is the Amazon, which is governed by multiple South American countries and inhabited by diverse groups of Indigenous Peoples. A series of negotiations at the United Nations is pushing to protect the high seas in a legally binding treaty—an ambitious move not unlike the signing of the Antarctic Treaty in the years following World War II.[6] And in March 2023, right before this book went to press, these negotiations succeeded with 193 countries reaching a historic agreement to protect the biodiversity of the high seas beyond national jurisdiction. Although the High Seas Treaty is not yet ratified, it lays the groundwork for sustainably using and protecting the ocean—from surface to seafloor—in a way that society has never granted the high seas before. This is the largest and most unknown habitat on Earth, one that has never received real support and protection or attracted the same amount of inquiry as outer space, even though it plays an outsized role in supporting life on Earth.

By the end of this decade, there might not be a complete map of the seafloor, but whenever Seabed 2030 is finished, I hope it leads the way to a new era in understanding and protecting the deep sea. The weird, wonderful deep-sea world is not a blank space, another frontier to use up and throw away. It is the last truly mysterious place on Earth, the birthplace of all life, an untapped medicine chest, a bulwark against climate chaos, and the key to discovering alien life beyond Earth. If we don't safeguard it for science, we will miss a historic opportunity to uncover our past and protect our future. This time, map in hand, I hope we find the right path.

ACKNOWLEDGMENTS

The Deepest Map began as I was procrastinating editing my last book. I read an article in *Smithsonian* magazine titled "Why the First Complete Map of the Ocean Floor Is Stirring Controversial Waters," and that 1,500-word piece by Kyle Frischkorn led to a few furiously creative hours, dreaming about what a book about mapping the seafloor might be. My first few interviews, with Vicki Ferrini at Lamont-Doherty Earth Observatory, Aileen Bohan at the Geological Survey Ireland, Jyotika Virmani and Carlie Wiener at the Schmidt Ocean Institute, and Luc Cuyvers at the Gallifrey Foundation, were incredibly insightful. An article I published in the *Guardian* soon afterward became the foundation of *The Deepest Map*—a huge thank-you to the Oceans editor, Chris Michael, for getting this project off the ground.

The story of two ships and their crews formed the backbone of *The Deepest Map*. First, a big thanks to the members of the Five Deeps Expedition, including Victor Vescovo, Alan Jamieson, Heather Stewart, Patrick Lahey, Rob McCallum, and, last but not least, Cassie Bongiovanni, who deserves my endless thanks for taking me on a crash course in ocean mapping over a couple of marathon Zoom calls. Next, I have to thank the crew aboard E/V *Nautilus*, in particular Nicole Raineault, who invited an untested landlubber on board for nine days and, most generously of all, shared her *private* cabin with me. What an angel. Thanks also to ocean mappers Renato Kane, Erin Heffron, and Anne Hartwell for talking, and sometimes taking, me through the ins and outs of mapping on board E/V *Nautilus*, and to Brooke Travis,

Sarah Lott, Michael Hannaford, Virginia Edgcomb, Kris Krasnosky, D. J. Yousavich, Tim Burbank, Christopher Powers, and Megan Cook for all their hospitality and help out at sea.

My profuse thanks to everyone whose stories, knowledge, and insights appeared in these pages: Rob Beaman, David Sandwell, Mike Coffin, Hyun-Chul Han, Jamie McMichael-Phillips, Helen Snaith, Philip Steinberg, Kevin Mackay, Larry Mayer, Richard Jenkins, Brian Connon, Tim Kearns, Julien Desrochers, Annamaria DeAngelis, Robin Falconer, Mitsuyuki Unno, John Hall, Jennifer Jencks, Shirley Tagalik, Kukik Baker, Joe Karetak, Andrew "Balum" Muckpah, Aupaa Irkok, Brian Calder, Mathieu Rondeau, Denis Hains, Shawn Joy, Morgan Smith, Michael Faught, Barbara Clark, Patricia Esquete, Karoline Postel-Vinay, and Arlo Hemphill.

Of course, many more people provided information and direction behind the scenes: Rika Anderson, Yves Guillam, Victoria Weda, Alissa Johnson, Lisa Levin, Andrew Friedman, Jacqueline Mammerickx, Bruce Strickrott, Jon Copley, Oliver Steeds, Lisa Hynes, Nicole Trenholm, Alice Doyle, Dan Fornari, Craig Young, Josh Young, Tomer and Ofer Ketter, Kirill Novoselskiy, Evert Flier, Diana Krawczyk, Andrew Goodwillie, Bob Fisher, Chris German, Jessi Halligan, Laurie Barge, Charles Kohnen, Peter Girguis, Guy "Harley" Means, Nick Bentley, Ximena Smith, Lindsay Gee, Debora Hansen Kleist, Karl Zinglersen, Ulrich Schwarz-Schampera, Hali Felt, Michael Smith, Lucas Owlijoot, Joe Shamee, Jacques John Mikiyungiak, and Jeff Kingston.

Many friends and colleagues read manuscripts and provided feedback at various stages: Ryan Hardy, Philip Steinberg, Ron Doel, Aletta Mondré, Starla Robinson, Rosemary Sullivan, and Helen Scales. My fellow board members at the San Diego Science Writing Association (Xochitl Rojas-Rocha, Monica May, Jared Whitlock, Ramin Skibba, Mario Aguilera, Patricia Fernandez, and Brittany Fair) cheered me on whenever I needed a boost. A special thanks to Michael Miller for shouldering part of my duties so I could meet my deadline in the final stages of writing.

Without the financial support of the Canada Council for the Arts, I could not have reported on Inuit hunters mapping their coastlines or on the ISA Council meetings in Jamaica. I am so grateful to be a citizen of a country that supports creators at all stages of their careers. I also received a generous grant from the Fund for Investigative Journalism, as well as legal advice from Jennifer Nelson and Celine Rohr at Reporters Committee for Freedom of the Press and fact-checking assistance from Emily Latimer.

Thank you to my agent, Suzy Evans, who believed in this book from the beginning and who then went out and found me incredible editors, Karen Rinaldi and Rachel Kambury at Harper Wave and Alan Sheppard at Goose Lane Editions.

Finally, I owe heartfelt thanks to my friends and family for supporting my work, especially my husband, who listened to me talk "shop" over dinner, and who read the manuscript and double-checked all my calculations because math scares me. I love you.

NOTES

PROLOGUE

1. "Seabed 2030 Announces Increase in Ocean Data Equating to the Size of Europe and Major New Partnership at UN Ocean Conference," The Nippon Foundation-GEBCO Seabed 2030 Project, https://seabed2030.org/news/sea bed-2030-announces-increase-ocean-data-equating-size-europe-and-major -new-partnership-un.

2. Dana Goodyear, "Without Sylvia Earle, We'd Be Living on Google Dirt," *New Yorker*, June 20, 2022, https://www.newyorker.com/magazine/2022/06 /27/without-sylvia-earle-wed-be-living-on-google-dirt.

3. Helen Scales, *The Brilliant Abyss: Exploring the Majestic Hidden Life of the Deep Ocean, and the Looming Threat That Imperils It* (New York: Atlantic Monthly Press, 2021), 4.

4. Casey Dreier, "The Cost of Perseverance, in Context," The Planetary Society, July 29, 2020, https://www.planetary.org/articles/cost-of-perseverance-in-context.

5. Ramin Skibba, "Why NASA Wants to Go Back to the Moon," *Wired*, August 12, 2022, https://www.wired.com/story/why-nasa-wants-to-go-back-to -the-moon/.

CHAPTER 1: AN EXPEDITION INTO THE DEEP

1. *The Ocean Economy in 2030* (Paris: OECD, 2016), https://doi.org/10.1787 /9789264251724-en.

2. Nicole Starosielski, *The Undersea Network* (Durham, NC: Duke University Press, 2015).

3. Josh Young, *Expedition Deep Ocean: The First Descent to the Bottom of the World's Oceans* (New York: Pegasus Books, 2020), 1–13.

4. Richard Mendick, "Richard Branson Abandons Ambitious Plan to Pilot Submarine to Deepest Points of Five Oceans," *National Post*, December 14, 2014, https://nationalpost.com/news/richard-branson-abandons-ambitious -plan-to-pilot-submarine-to-deepest-points-of-five-oceans.

5. John Nelson, "How Deep Is Challenger Deep?," ArcGIS StoryMaps, August 3, 2020, https://storymaps.arcgis.com/stories/0d389600f3464e318 5a84c199f04e859.

6. Young, *Expedition Deep Ocean*, 20.

7. David Grann, *The Lost City of Z: A Legendary British Explorer's Deadly Quest to Uncover the Secrets of the Amazon* (London: Simon & Schuster, 2017), 58.

8. Ben Taub, "Thirty-Six Thousand Feet Under the Sea," *New Yorker*, May 2020, https://www.newyorker.com/magazine/2020/05/18/thirty-six-thousand-feet-under-the-sea.

9. "New Dives to Challenger Deep Raise Old Questions About Privatization and Exploration," *DSM Observer*, July 21, 2020, https://dsmobserver.com/2020/07/new-dives-to-challenger-deep-raise-old-questions-about-privatization-and-exploration/.

10. Anne-Cathrin Wölfl et al., "Seafloor Mapping—The Challenge of a Truly Global Ocean Bathymetry," *Frontiers in Marine Science* 6 (June 5, 2019): 283, https://doi.org/10.3389/fmars.2019.00283.

11. Robert Kunzig, *Mapping the Deep: The Extraordinary Story of Ocean Science* (New York: W. W. Norton, 2000), 65.

12. Helen Scales, *The Brilliant Abyss: Exploring the Majestic Hidden Life of the Deep Ocean, and the Looming Threat That Imperils It* (New York: Atlantic Monthly Press, 2021), 5.

13. Kunzig, *Mapping the Deep*, 59–60.

14. Jacqueline Carpine-Lancre et al., *History of GEBCO: 1903–2003* (Utrecht, Netherlands: GITC by Lemmer, 2003), 1.

15. David T. Sandwell et al., "New Global Marine Gravity Model from CryoSat-2 and Jason-1 Reveals Buried Tectonic Structure," *Science* 346, no. 6205 (October 3, 2014): 65–67, https://doi.org/10.1126/science.1258213.

16. Jon Copley, "Just How Little Do We Know About the Ocean Floor?," *The Conversation*, October 9, 2014, https://theconversation.com/just-how-little-do-we-know-about-the-ocean-floor-32751.

17. John Noble Wilford, *The Mapmakers: The Story of the Great Pioneers in Cartography from Antiquity to the Space Age* (New York: Knopf, 1981), 328.

18. Young, *Expedition Deep Ocean*, 98.

19. Ibid., 88.

20. "James Cameron: Diving Deep, Dredging Up Titanic," NPR, March 30, 2012, https://www.npr.org/2012/03/30/149635287/james-cameron-diving-deep-dredging-up-titanic; "Director James Cameron Reveals He Directs Movies Just to Make Money for Deep Sea Exploration," *Daily Telegraph*, May 29, 2018, https://dailytelegraph.com.au/entertainment/sydney-confidential/director-james-cameron-reveals-he-directs-movies-just-to-make-money-for-deep-sea-exploration/news-story/d5380ef4ec58883bf0ecf8cdde40da96.

21. Taub, "Thirty-Six Thousand Feet Under the Sea."

22. Young, *Expedition Deep Ocean*, 52–62.

CHAPTER 2: LOOKING FOR A SHIP

1. Anne-Cathrin Wölfl et al., "Seafloor Mapping—The Challenge of a Truly Global Ocean Bathymetry," *Frontiers in Marine Science* 6 (June 5, 2019): 283, https://doi.org/10.3389/fmars.2019.00283.

2. Alan J. Jamieson et al., "Fear and Loathing of the Deep Ocean: Why Don't People Care About the Deep Sea?," *ICES Journal of Marine Science* 78, no. 3 (July 2021), https://doi.org/10.1093/icesjms/fsaa234.

3. The Nippon Foundation-GEBCO Seabed 2030 Project, "Deep Ambition: How to Map the World," 2020.

4. Alan J. Jamieson and Thomas Linley, hosts, "The Moon Analogy. Guest: Monty Priede," *The Deep-Sea Podcast*, episode 001, Armatus Oceanic, July 8, 2020, https://www.armatusoceanic.com/podcast/episode1.

5. K. Picard, B. Brooke, and M. F. Coffin, "Geological Insights from Malaysia Airlines Flight MH370 Search," *Eos*, March 6, 2017, https://eos.org /science-updates/geological-insights-from-malaysia-airlines-flight-mh370 -search; Sarah Zhang, "The Search for MH370 Revealed Secrets of the Deep Ocean," *Atlantic*, March 10, 2017, https://www.theatlantic.com/science/archive /2017/03/mh370-search-ocean/518946/.

6. International Hydrographic Organization, *Measuring and Charting the Ocean: One Hundred Years of International Cooperation in Hydrography* (Hamburg, Germany: International Hydrographic Organization, March 2020), 15, https://iho. int/publications.

7. NOAA, "NOAA Research—Budget 2022," 2020, 9, https://research.noaa .gov/External-Affairs/Budget.

8. Brian Dunbar, "FY 2021 NASA Budget," NASA, June 30, 2022, https://www .nasa.gov/content/fy-2021-nasa-budget.

9. "President Trump's Bold Vision Will Help Conserve, Manage, and Explore America's Oceans," White House, January 5, 2021, https://www.whitehouse .gov/articles/president-trumps-bold-vision-will-help-conserve-manage-explore -americas-oceans/.

10. Evan Lubofsky, "The Discovery of Hydrothermal Vents," *Oceanus*, June 11, 2018, https://www.whoi.edu/oceanus/feature/the-discovery-of-hydrothermal -vents/.

11. William J. Broad, "*Titanic* Wreck Was Surprise Yield of Underwater Tests," *New York Times*, September 8, 1985, https://www.nytimes.com/1985/09/08 /us/titanic-wreck-was-surrise-yield-of-underwater-tests-for-military.html.

12. Eric Levenson, "Inside the Secret US Military Mission That Located the *Titanic*," CNN, December 13, 2018, https://www.cnn.com/2018/12/13/us /titanic-discovery-classified-nuclear-sub/index.html.

13. Donald J. Trump, "Memorandum on Ocean Mapping of the United States Exclusive Economic Zone and the Shoreline and Nearshore of Alaska," White

House, November 19, 2019, https://trumpwhitehouse.archives.gov
/presidential-actions/memorandum-ocean-mapping-united-states-exclusive
-economic-zone-shoreline-nearshore-alaska/.

14. "Read the Rainbow: Seafloor Mapping Glossary," Nautilus Live, August 10,
2018, https://nautiluslive.org/blog/2018/08/10/read-rainbow-seafloor-map
ping-glossary.

15. "The Mysterious 'False Bottom' of the Twilight Zone," Woods Hole Oceanographic Institution, April 26, 2022, https://twilightzone.whoi.edu/the-myster
ious-false-bottom-of-the-twilight-zone/.

16. Natacha Aguilar de Soto et al., "Anthropogenic Noise Causes Body Malforma-
tions and Delays Development in Marine Larvae," *Scientific Reports* 3 (October 3, 2013): article 2831, https://doi.org/10.1038/srep02831.

17. Sophie L. Nedelec et al., "Anthropogenic Noise Playback Impairs Embryonic
Development and Increases Mortality in a Marine Invertebrate," *Scientific
Reports* 4 (July 31, 2014): article 5891, https://doi.org/10.1038/srep05891.

18. Ian T. Jones, Jenni A. Stanley, and T. Aran Mooney, "Impulsive Pile Driving
Noise Elicits Alarm Responses in Squid (*Doryteuthis Pealeii*)," *Marine Pollution Bulletin* 150 (January 2020): article 110792, https://doi.org/10.1016/j.mar
polbul.2019.110792.

19. Joy E. Stanistreet et al., "Changes in the Acoustic Activity of Beaked Whales
and Sperm Whales Recorded During a Naval Training Exercise off Eastern
Canada," *Scientific Reports* 12, no. 1 (February 7, 2022): article 1973, https://
doi.org/10.1038/s41598-022-05930-4.

20. Anne E. Simonis et al., "Co-occurrence of Beaked Whale Strandings and
Naval Sonar in the Mariana Islands, Western Pacific," *Proceedings of the
Royal Society B: Biological Sciences* 287, no. 1921 (February 26, 2020): article
20200070, https://doi.org/10.1098/rspb.2020.0070.

CHAPTER 3: TO THE BOTTOM OF THE ATLANTIC

1. Josh Young, *Expedition Deep Ocean: The First Descent to the Bottom of the
World's Oceans* (New York: Pegasus Books, 2020), 52–64.

2. "Ocean's Deepest Point Conquered," Guinness World Records channel on
YouTube, November 24, 2020, https://www.youtube.com/watch?v=ulIQ9
_BB8KA.

3. Ben Taub, "Thirty-Six Thousand Feet Under the Sea," *New Yorker*, May 2020,
https://www.newyorker.com/magazine/2020/05/18/thirty-six-thousand-feet
-under-the-sea.

4. Heather A. Stewart and Alan J. Jamieson, "The Five Deeps: The Location and
Depth of the Deepest Place in Each of the World's Oceans," *Earth-Science
Reviews* 197 (October 2019): 5, https://doi.org/10.1016/j.earscirev
.2019.102896.

5. Cassandra Bongiovanni, Heather A. Stewart, and Alan J. Jamieson, "High-Resolution Multibeam Sonar Bathymetry of the Deepest Place in Each Ocean," *Geoscience Data Journal* 9, no. 1 (June 2022): 108–122, https://doi.org/10.1002/gdj3.122.

6. Young, *Expedition Deep Ocean*, 140–41.

7. Larry Mayer, "UN Decade of Ocean Science," Map the Gaps Symposium, Paris, January 11, 2021, https://mapthegapssymposium2021.sched.com/event/gUx7/mtg-symposium-un-decade-of-ocean-science?iframe=no.

8. William J. Broad, "So You Think You Dove the Deepest? James Cameron Doesn't," *New York Times*, September 16, 2019, https://www.nytimes.com/2019/09/16/science/ocean-sea-challenger-exploration-james-cameron.html.

9. Taub, "Thirty-Six Thousand Feet Under the Sea."

10. Young, *Expedition Deep Ocean*, xiv.

11. Kelsey Kennedy, "The Forgotten Documents of a 1918 Tsunami in Puerto Rico," *Atlas Obscura*, July 5, 2017, https://www.atlasobscura.com/articles/puerto-rico-earthquake-tsunami-lost-records.

12. *Expedition Deep Ocean* (Discovery Channel, 2021), https://www.discoveryplus.com/show/expedition-deep-ocean.

13. Helen Scales, *The Brilliant Abyss: Exploring the Majestic Hidden Life of the Deep Ocean, and the Looming Threat That Imperils It* (New York: Atlantic Monthly Press, 2021), 4.

14. Robert Ballard, "The Astonishing Hidden World of the Deep Ocean," transcript, TED Talk, Monterey, California, May 2008, https://www.ted.com/talks/robert_ballard_the_astonishing_hidden_world_of_the_deep_ocean/transcript.

15. "Prince of Monaco Here on His Yacht," *New York Times*, September 11, 1913.

16. Robert Kunzig, *Mapping the Deep: The Extraordinary Story of Ocean Science* (New York: W. W. Norton, 2000), 276.

17. International Hydrographic Organization, *Measuring and Charting the Ocean: One Hundred Years of International Cooperation in Hydrography* (Hamburg, Germany: International Hydrographic Organization, March 2020), 27, https://iho.int/publications.

18. Lloyd A. Brown, *The Story of Maps* (Boston: Little, Brown, 1949), 144.

19. Jacqueline Carpine-Lancre et al., *History of GEBCO: 1903–2003* (Utrecht, Netherlands: GITC by Lemmer, 2003), 13.

20. David E. Kaplan and Alec Dubro, *Yakuza: The Explosive Account of Japan's Criminal Underworld* (San Francisco: Center for Investigative Reporting, 1986), 79.

21. Karoline Postel-Vinay with Mark Selden, "History on Trial: French Nippon Foundation Sues Scholar for Libel to Protect the Honor of Sasakawa Ryo-ichi,"

Asia-Pacific Journal: Japan Focus 8, no. 17 (April 26, 2010): article 3349, https://apjjf.org/-Mark-Selden/3349/article.html.

22. "Obituary: Ryoichi Sasakawa," *Independent*, July 19, 1995, https://www.indep endent.co.uk/news/people/obituary-ryoichi-sasakawa-1592324.html; *Anne-Marie Sauteraud*, case no. 09/04019, Tribunal de Grande Instance de Paris, September 22, 2010.

23. "The Godfather-san," *Time*, August 26, 1974, https://content.time.com/time /subscriber/article/0,33009,944948-1,00.html.

24. Kaplan and Dubro, *Yakuza*, 261–62.

25. Postel-Vinay with Selden, "History on Trial."

26. Lisa Torio, "Abe's Japan Is a Racist, Patriarchal Dream," Jacobin, March 28, 2017, https://jacobin.com/2017/03/abe-nippon-kaigi-japan-far-right/; Sachie Mizohata, "Nippon Kaigi: Empire, Contradiction, and Japan's Future," *Asia-Pacific Journal: Japan Focus* 14, no. 21 (November 1, 2016): article 4975, https://apjjf.org/2016/21/Mizohata.html.

27. *Fondation Franco-Japonese Sasakawa vs. Karoline Postel-Vinay*, Tribunal de Grande Instance de Paris September 22, 2010.

28. Ibid.

29. Jeff Kingston, "Japanese Revisionists' Meddling Backfires," *Critical Asian Studies* 51, no. 3 (June 23, 2019): 437–50, https://doi.org/10.1080/14672715 .2019.1627889.

30. "Obituary: Ryoichi Sasakawa," *Independent*, July 19, 1995, https://www.indep endent.co.uk/news/people/obituary-ryoichi-sasakawa-1592324.html.

31. Heather A. Stewart and Alan J. Jamieson, "The Five Deeps: The Location and Depth of the Deepest Place in Each of the World's Oceans," *Earth-Science Reviews* 197 (October 2019): article 102896, https://doi.org/10.1016 /j.earscirev.2019.102896.

32. "What Are the Roaring Forties?," National Ocean Service, October 25, 2020, https://oceanservice.noaa.gov/facts/roaring-forties.html.

CHAPTER 4: MARIE THARP AND THE MAP THAT CHANGED THE WORLD

1. Henry David Thoreau, *Cape Cod* (1865; repr. New York: Thomas Y. Crowell & Co., 1908), 141.

2. Susan Schulten, *A History of America in 100 Maps* (Chicago: University of Chicago Press, 2018), 262.

3. John Noble Wilford, *The Mapmakers: The Story of the Great Pioneers in Cartography from Antiquity to the Space Age* (New York: Knopf, 1981), 280.

4. Hali Felt, *Soundings: The Story of the Remarkable Woman Who Mapped the Ocean Floor* (New York: Henry Holt, 2012), 273.

5. Marie Tharp, "Connect the Dots: Mapping the Seafloor and Discovering the

Mid-ocean Ridge," in *Lamont-Doherty Earth Observatory of Columbia: Twelve Perspectives on the First Fifty Years, 1949–1999*, edited by Laurence Lippsett (Palisades, NY: Lamont-Doherty Earth Observatory, 1999), chapter 2, https://www.whoi.edu/news-insights/content/marie-tharp/.

6. Interview of Marie Tharp by Ronald Doel, Session I, September 14, 1994, Niels Bohr Library & Archives, American Institute of Physics, College Park, Maryland (hereafter AIP), https://www.aip.org/history-programs/niels-bohr-library/oral-histories/6940.

7. Tharp, "Connect the Dots."

8. Naomi Oreskes, *Science on a Mission: How Military Funding Shaped What We Do and Don't Know About the Ocean* (Chicago: University of Chicago Press, 2021), 262.

9. Interview of Tharp by Doel, Session I, September 14, 1994, AIP.

10. Tharp, "Connect the Dots."

11. Interview of W. Arnold Finck by Ronald Doel, Session I, March 11, 1996, AIP, https://www.aip.org/history-programs/niels-bohr-library/oral-histories/6948-1.

12. Interview of Alma Kesner by Ronald Doel, Session I, October 25, 1995, AIP, https://www.aip.org/history-programs/niels-bohr-library/oral-histories/6947-1.

13. Ronald E. Doel, Tanya J. Levin, and Mason K. Marker, "Extending Modern Cartography to the Ocean Depths: Military Patronage, Cold War Priorities, and the Heezen-Tharp Mapping Project, 1952–1959," *Journal of Historical Geography* 32, no. 3 (July 2006): 610, https://doi.org/10.1016/j.jhg.2005.10.011.

14. Robert Kunzig, *Mapping the Deep: The Extraordinary Story of Ocean Science* (New York: W. W. Norton, 2000), 58.

15. Bruce Heezen, Marie Tharp, and William Ewing, *The Floors of the Oceans: I. The North Atlantic*, Special Paper 65 (New York: Geological Society of America, 1959), https://www.gutenberg.org/files/49069/49069-h/49069-h.htm#Page_3.

16. Kunzig, *Mapping the Deep*, 40–41.

17. Tharp, "Connect the Dots."

18. Interview of Kesner by Doel, Session I, October 25, 1995, AIP.

19. Suzanne O'Connell, "Marie Tharp Pioneered Mapping the Bottom of the Ocean 6 Decades Ago—Scientists Are Still Learning about Earth's Last Frontier," *The Conversation*, July 28, 2020.

20. Tharp, "Connect the Dots."

21. Henry William Menard, *The Ocean of Truth: A Personal History of Global Tectonics* (Princeton, NJ: Princeton University Press, 1986), 26.

22. Andrea Wulf, *The Invention of Nature: Alexander von Humboldt's New World* (New York: Vintage Books, 2015), 4.

23. Menard, *The Ocean of Truth*, 20.

24. Naomi Oreskes, *Plate Tectonics: An Insider's History of the Modern Theory of the Earth* (Boulder, CO: Westview Press, 2001), 7–12.

25. Menard, *The Ocean of Truth*, 27.

26. Kunzig, *Mapping the Deep*, 33.

27. Helen M. Rozwadowski, *Fathoming the Ocean: The Discovery and Exploration of the Deep Sea* (Cambridge, MA: Harvard University Press, 2005), 30.

28. Stephen Dowling, "The Quest That Discovered Thousands of New Species," BBC Future, February 5, 2021, https://www.bbc.com/future/article/2021 0204-the-quest-that-discovered-thousands-of-new-species.

29. Rozwadowski, *Fathoming the Ocean*, 62.

30. Kunzig, *Mapping the Deep*, 32–38.

31. Rosalind Williams, *Notes on the Underground: An Essay on Technology, Society, and the Imagination* (Cambridge, MA: MIT Press, 1990), 193.

32. Menard, *The Ocean of Truth*, 21.

33. Interview of Alma Kesner by Ronald Doel, Session II, May 18, 1997, AIP, https://www.aip.org/history-programs/niels-bohr-library/oral-histories/6947-2.

34. Enrico Bonatti and Kathleen Crane, "Oceanography and Women: Early Challenges," *Oceanography* 25, no. 4 (December 2012): 33, https://doi.org /10.5670/oceanog.2012.103.

35. Interview of Kesner by Doel, Session I, October 25, 1995, AIP.

36. Menard, *The Ocean of Truth*, 42.

37. Bonatti and Crane, "Oceanography and Women," 37.

38. Oreskes, *Science on a Mission*, 244.

39. Menard, *The Ocean of Truth*.

40. Tharp, "Connect the Dots."

41. Menard, *The Ocean of Truth*, 107.

42. Interview of Marie Tharp by Ronald Doel, Session II, December 18, 1996, AIP, https://www.aip.org/history-programs/niels-bohr-library/oral-histories /22896-2.

43. Bonatti and Crane, "Oceanography and Women," 32–39.

44. Ibid., 37.

45. Interview of Marie Tharp by Tanya Levin, Session IV, June 28, 1997, AIP, https://www.aip.org/history-programs/niels-bohr-library/oral-histories/22 896-4.

46. Menard, *The Ocean of Truth*, 61.

47. Schulten, *A History of America in 100 Maps*, 18.

48. Ibid., 118.

49. Marie DeNoia Aronsohn, "Lamont's Marie Tharp: She Drew the Maps That Shook the World," Columbia Climate School, July 27, 2020, https://news.climate.columbia.edu/2020/07/27/marie-tharp-maps-legacy/.

50. Interview of Kesner by Doel, Session II, May 18, 1997, AIP.

51. Laurie Lawlor, *Super Women: Six Scientists Who Changed the World* (New York City: Holiday House Publishing, 2017).

52. Menard, *The Ocean of Truth*, 29.

53. Tharp, "Connect the Dots."

54. Interview of Tharp by Levin, Session IV, June 28, 1997, AIP. https://www.aip.org/history-programs/niels-bohr-library/oral-histories/22896-4.

55. Tharp, "Connect the Dots."

56. Menard, *The Ocean of Truth*, 94–95.

57. Interview of Tharp by Levin, Session IV, June 28, 1997, AIP.

58. Oreskes, *Plate Tectonics*, xx.

59. "Pioneers of Plate Tectonics: John Tuzo-Wilson," Geological Society of London, https://www.geolsoc.org.uk/Plate-Tectonics/Chap1-Pioneers-of-Plate-Tectonics/John-Tuzo-Wilson.

60. Ken MacDonald, "What Is the Mid-ocean Ridge?," National Oceanic and Atmospheric Administration, https://oceanexplorer.noaa.gov/explorations/05galapagos/background/mid_ocean_ridge/mid_ocean_ridge.html.

61. Oreskes, *Plate Tectonics*, xi–xx.

62. Heezen, Tharp, and Ewing, *The Floors of the Oceans*.

63. Interview of Tharp by Levin, Session IV, June 28, 1997, AIP.

64. Kunzig, *Mapping the Deep*, 62–63.

65. Simon Winchester, *Land: How the Hunger for Ownership Shaped the Modern World* (New York: Harper, 2021).

66. Stephen Hall, *Mapping the Next Millennium: How Computer-Driven Cartography Is Revolutionizing the Face of Science* (New York: Random House, 1992).

67. Interview of Tharp by Doel, Session I, September 14, 1994, AIP.

68. Interview of Marie Tharp by Tanya Levin, Session III, May 24, 1997, AIP, https://www.aip.org/history-programs/niels-bohr-library/oral-histories/22896-3.

69. Interview of Tharp by Levin, Session IV, June 28, 1997, AIP.

70. Tharp, "Connect the Dots."

71. Interview of Kesner by Doel, Session II, May 18, 1997, AIP.

72. Menard, *The Ocean of Truth*, 199.

73. Ibid.

74. Ibid., 199–200.

75. Ibid., 201.

76. Interview of Kesner by Doel, Session II, May 18, 1997, AIP.

77. Ibid.

78. Ibid.

79. Menard, *The Ocean of Truth*, 199.

80. Kunzig, *Mapping the Deep*, 56.

81. Interview of Kesner by Doel, Session II, May 18, 1997, AIP.

82. Ibid.

83. Robert Ballard, "The Astonishing Hidden World of the Deep Ocean," transcript, TED Talk, Monterey, California, May 2008, https://www.ted.com /talks/robert_ballard_the_astonishing_hidden_world_of_the_deep_ocean /transcript.

84. Schulten, *A History of America in 100 Maps*, 12–14.

85. Interview of Tharp by Doel, Session I, September 14, 1994, AIP.

86. Valerie J. Nelson, "Marie Tharp, 86; Pioneering Maps Altered Views on Seafloor Geology," *Los Angeles Times*, September 4, 2006, https://www.latimes .com/archives/la-xpm-2006-sep-04-me-tharp4-story.html.

CHAPTER 5: THE LONELIEST OCEAN ON EARTH

1. Josh Young, *Expedition Deep Ocean: The First Descent to the Bottom of the World's Oceans* (New York: Pegasus Books, 2020), 170.

2. Sarah Gibbens, "There's a New Ocean Now—Can You Name All 5?," *National Geographic*, August 6, 2021, https://www.nationalgeographic.com/environment /article/theres-a-new-ocean-now-can-you-name-all-five-southern-ocean.

3. Derek Lundy, *Godforsaken Sea: The True Story of a Race Through the World's Most Dangerous Waters* (Chapel Hill, NC: Algonquin Books of Chapel Hill, 1998), 5.

4. Young, *Expedition Deep Ocean*, 175.

5. Cassandra Bongiovanni, Heather A. Stewart, and Alan J. Jamieson, "High-Resolution Multibeam Sonar Bathymetry of the Deepest Place in Each Ocean," *Geoscience Data Journal* 9, no. 1 (June 2022): 108–23, https://doi .org/10.1002/gdj3.122.

6. "Licence to Krill," Greenpeace International, March 12, 2018, https://www .greenpeace.org/international/publication/15255/licence-to-krill-antarctic-krill -report.

7. Kendall R. Jones et al., "The Location and Protection Status of Earth's Diminishing Marine Wilderness," *Current Biology* 28, no. 15 (August 6, 2018): 2506–12.E3, https://doi.org/10.1016/j.cub.2018.06.010.

8. Young, *Expedition Deep Ocean*, 179–80.

9. Ibid., 184.

10. Ibid., 184–85.

11. Helen Scales, *The Brilliant Abyss: Exploring the Majestic Hidden Life of the Deep Ocean, and the Looming Threat That Imperils It* (New York: Atlantic Monthly Press, 2021), 21.

12. "South Sandwich Trench," Wikipedia, November 10, 2020, https://en.wiki pedia.org/w/index.php?title=South_Sandwich_Trench&oldid=988015061.

13. Young, *Expedition Deep Ocean*, 186.

14. Victor Vescovo, "Southern Ocean Expedition Blog," The Five Deeps Expedition, February 22, 2019, https://fivedeeps.com/home/expedition/southern/live/.

15. Lloyd A. Brown, *The Story of Maps* (Boston: Little, Brown, 1949), 149.

16. Brad Lendon, "Analysis: How Did a $3 Billion US Navy Submarine Hit an Undersea Mountain?," CNN, November 4, 2021, https://www.cnn.com /2021/11/04/asia/submarine-uss-connecticut-accident-undersea-mountain -hnk-intl-ml-dst/index.html.

17. Five Deeps, "Naming," The Five Deeps Expedition, https://fivedeeps.com /home/technology/names/.

18. Michael Huet, "International Naming of Undersea Features," GEBCO Sub-Committee on Undersea Feature Names, n.d.

19. Simon Winchester, *Land: How the Hunger for Ownership Shaped the Modern World* (New York: Harper, 2021).

CHAPTER 6: NAMING AND CLAIMING THE SEAFLOOR

1. "Treaty of Waitangi," New Zealand Ministry of Justice, March 11, 2020, https://www.justice.govt.nz/about/learn-about-the-justice-system/how-the -justice-system-works/the-basis-for-all-law/treaty-of-waitangi/.

2. Jacqueline Carpine-Lancre et al., *History of GEBCO: 1903–2003* (Utrecht, Netherlands: GITC by Lemmer, 2003), 107.

3. Ibid., 109.

4. J. Brian Harley, "Maps, Knowledge, and Power," in *Geographic Thought : A Praxis Perspective*, edited by George L. Henderson and Marvin Waterstone (London: Routledge, 2009), 134–35.

5. Bill Hayton, *The South China Sea: The Struggle for Power in Asia* (New Haven, CT: Yale University Press, 2014), 92–93.

6. Carpine-Lancre et al., *History of GEBCO*, 109.

7. The Nippon Foundation-GEBCO Seabed 2030 Project, "Deep Ambition: How to Map the World," 2020.

8. Tegg Westbrook, "The Global Positioning System and Military Jamming: Geographies of Electronic Warfare," *Journal of Strategic Security* 12, no. 2 (2019): 1–2.

9. Bill Hayton, "The South China Sea in 2020: Statement before the

U.S.-China Economic and Security Review Commission Hearing on 'U.S.-China Relations in 2020: Enduring Problems and Emerging Challenges,'" U.S.-China Economic and Security Review Commission, September 9, 2020, 3, https://www.uscc.gov/sites/default/files/2020-09/Hayton _Testimony.pdf.

10. Ivan Watson, Brad Lendon, and Ben Westcott, "Inside the Battle for the South China Sea," CNN, August 2018, https://www.cnn.com/interactive /2018/08/asia/south-china-sea/.

11. Luc Cuyvers et al., *Deep Seabed Mining: A Rising Environmental Challenge* (Gland, Switzerland: IUCN and Gallifrey Foundation, 2018), 32.

12. Vo Kieu Bao Uyen and Shashank Bengali, "Sunken Boats. Stolen Gear. Fishermen Are Prey as China Conquers a Strategic Sea," *Los Angeles Times*, November 12, 2020, https://www.latimes.com/world-nation/story/2020-11-12 /china-attacks-fishing-boats-in-conquest-of-south-china-sea.

13. Hayton, *The South China Sea*, 113.

14. Max Fisher, "The South China Sea: Explaining the Dispute," *New York Times*, July 14, 2016, https://www.nytimes.com/2016/07/15/world/asia/south -china-sea-dispute-arbitration-explained.html.

15. Hayton, "The South China Sea in 2020," 2.

16. Zachery Haver, "China Trademarked Hundreds of South China Sea Landmarks," BenarNews, April 13, 2021, https://www.benarnews.org/english /news/philippine/sea-trademarks-04132021172405.html.

17. Yukie Yoshikawa, "The US-Japan-China Mistrust Spiral and Okinotorishima," *Asia-Pacific Journal* 5, no. 10 (October 1, 2007): article 2541, https://apjjf.org/- Yukie-YOSHIKAWA/2541/article.html.

18. Norimitsu Onishi, "Japan and China Dispute a Pacific Islet," *New York Times*, July 10, 2005, https://www.nytimes.com/2005/07/10/world/asia/japan-and -china-dispute-a-pacific-islet.html.

19. Hayton, *The South China Sea*, 262.

20. Sung Hyo Hyun, "The Geomorphic Characteristics and Naming of Undersea Feature in the East Sea, Korea," in *The 14th International Seminar on Sea Names Geography, Sea Names, and Undersea Feature Names* (Tunis Ville, Tunisia: Society for East Sea, 2008).

21. F. Pappalardi, S. J. Dunham, and M. E. Leblang, "HMS Scott—United Kingdom Ocean Survey Ship," in *OCEANS 2000 MTS/IEEE Conference and Exhibition, Conference Proceedings*, vol. 2 (Providence, Rhode Island: IEEE, 2000), 961–67, https://doi.org/10.1109/OCEANS.2000.881724.

22. Stephen Hall, *Mapping the Next Millennium: The Discovery of New Geographies* (New York: Random House, 1992), 79–81.

23. Robert Kunzig, *Mapping the Deep: The Extraordinary Story of Ocean Science* (New York: W. W. Norton, 2000), 65.

24. John K. Hall, "Insider's View: Arctic Low-Budget Hydrography Update," *Hydro*

International, May 2008, https://www.hydro-international.com/content
/article/arctic-low-budget-hydrography-update.

25. "Safety & Shipping Review 2021," Allianz Global Corporate & Specialty, August 2021, 9, https://www.agcs.allianz.com/news-and-insights/reports/shipping-safety/shipping-report.html.

CHAPTER 7: CROWDSOURCING A MAP OF THE ARCTIC

1. Heather A. Stewart and Alan J. Jamieson, "The Five Deeps: The Location and Depth of the Deepest Place in Each of the World's Oceans," *Earth-Science Reviews* 197 (October 2019): article 102896, https://doi.org/10.1016/j.earscirev.2019.102896.

2. "Facts and Figures," Port of Rotterdam, https://www.portofrotterdam.com/en/experience-online/facts-and-figures.

3. "Innovative Hydrography," Port of Rotterdam, https://www.hydro-international.com/content/article/port-of-rotterdam-innovative-hydrography.

4. "About Our Hydrographic Service," Port of London Authority, https://www.pla.co.uk/Safety/About-Our-Hydrographic-Service.

5. Fisheries and Oceans Canada, "Arctic Charting," October 3, 2022, Government of Canada, https://charts.gc.ca/arctic-arctique/index-eng.html.

6. R. Glenn Wright and Michael Baldauf, "Arctic Environment Preservation Through Grounding Avoidance," in *Sustainable Shipping in a Changing Arctic*, edited by Lawrence P. Hildebrand, Lawson W. Brigham, and Tafsir M. Johansson, vol. 7, *WMU Studies in Maritime Affairs* (Cham, Switzerland: Springer International Publishing, 2018), 77, https://doi.org/10.1007/978-3-319-78425-0_5.

7. Heïdi Sevestre, "Life in One of the Fastest-Warming Places on Earth," Arctic Council, May 10, 2010, https://arctic-council.org/news/life-in-one-of-the-fastest-warming-places-on-earth/.

8. Lara Johannsdottir, David Cook, and Gisele M. Arruda, "Systemic Risk of Cruise Ship Incidents from an Arctic and Insurance Perspective," *Elementa: Science of the Anthropocene* 9, no. 1 (2021): article 00009, https://doi.org/10.1525/elementa.2020.00009.

9. Jackie Dawson et al., "Temporal and Spatial Patterns of Ship Traffic in the Canadian Arctic from 1990 to 2015," *Arctic* 71, no. 1 (2018): 15–26.

10. Wright and Baldauf, "Arctic Environment Preservation Through Grounding Avoidance."

11. Ibid., 90.

12. Grant Sims, "A Clot in the Heart of the Earth," *Outside*, June 1989, https://www.outsideonline.com/adventure-travel/clot-heart-earth/.

13. Karen Nasmith and Michael Sullivan, "Climate Change Adaptation Action Plan for Hamlet of Arviat" (Ottawa, Ontario: Canadian Institute of Planners,

July 2010), 10, https://www.climatechangenunavut.ca/sites/default/files/arviat
_community_adap_plan_eng.pdf.

14. Joshua Rapp Learn, "Arctic Search-and-Rescue Missions Double as Climate
    Warms," *National Geographic*, September 19, 2016, https://www.nationalgeo
    graphic.com/adventure/article/arctic-search-and-rescue-missions-double.

15. "Focus on Geography Series, 2016 Census: Arviat, Hamlet (CSD)—Nunavut,"
    Statistics Canada, https://www12.statcan.gc.ca/census-recensement/2016/as-sa
    /fogs-spg/Facts-csd-eng.cfm?LANG=Eng&GK=CSD&GC=6205015&
    TOPIC=4.

16. "RCMP Charge 30-Year-Old After Fatal Hit and Run," *Nunatsiaq News*, Sep-
    tember 27, 2021, https://nunatsiaq.com/stories/article/man-charged-follow
    ing-fatal-rankin-inlet-hit-and-run/.

17. Cheryl Katz, "With Old Traditions and New Tech, Young Inuit Chart Their
    Changing Landscape," *Hakai* magazine, August 30, 2022, https://hakaimaga
    zine.com/features/with-old-traditions-and-new-tech-young-inuit-chart-their
    -changing-landscape/.

18. Nickita Longman, "Hunger in the North," *University of Toronto Magazine*, Sep-
    tember 23, 2021, https://magazine.utoronto.ca/research-ideas/health
    /hunger-in-the-north-arviat-nunavut-food-insecurity/.

19. Emma Tranter, "Nunavut Children Experience the Highest Poverty Rate in
    Canada: Report," *Nunatsiaq News*, January 30, 2020, https://nunatsiaq.com
    /stories/article/nunavut-children-experience-the-highest-poverty-rate-in-canada
    -report/.

20. Jane George, "Tiny Homes Could Cure Western Nunavut Town's Growing
    Pains," *Nunatsiaq News*, September 20, 2016, https://nunatsiaq.com/stories
    /article/65674tiny_homes_could_cure_western_nunavut_towns_growing
    _pains/.

21. Kitra Cahana and Ed Ou, "How Teen Dance Competitions Are Helping Nun-
    avut Youth Fight Suicide," CBC News, https://www.cbc.ca/news2/inter
    actives/arviat-documentary/http://www.cbcnews.ca/teendance; "How a Dance
    Competition Helps Keep Suicide at Bay | Dancing Towards the Light," CBC
    News channel on YouTube, May 16, 2017, https://www.youtube.com/watch
    ?v=BZUwB-aNYp8.

22. Helen Epstein, "The Highest Suicide Rate in the World," *New York Review of
    Books*, https://www.nybooks.com/articles/2019/10/10/inuit-highest-suicide
    -rate/.

23. Peter Varga, "Arviat Fishermen Found Dead After Apparent Boating Acci-
    dent," *Nunatsiaq News*, August 12, 2014, https://nunatsiaq.com/stories/article
    /65674arviat_fishermen_found_dead_after_apparent_boating_accident/.

24. Learn, "Arctic Search-and-Rescue Missions Double as Climate Warms."

25. Katz, "With Old Traditions and New Tech, Young Inuit Chart Their Changing
    Landscape."

26. Jake Eggleston and Jason Pope, "Land Subsidence and Relative Sea-Level Rise

in the Southern Chesapeake Bay Region," Circular, Circular (U.S. Geological Survey Circular 1392, 2013), 2, https://dx.doi.org/10.3133/cir1392.

27. Pia Blake, "Mapping Future Canadian Arctic Coastlines," BA thesis, Lund University, Lund, Sweden, 2021, https://lup.lub.lu.se/student-papers/record /9056990.

28. Jill Barber, "Carving Out a Future: Contemporary Inuit Sculpture of Third Generation Artists from Arviat, Cape Dorset and Clyde River," MA dissertation, Carleton University, Ottawa, Ontario, 1999, 28–30.

29. Frédéric Laugrand, Jarich Oosten, and David Serkoak, "'The Saddest Time of My Life': Relocating the Ahiarmiut from Ennadai Lake (1950–1958)," *Polar Record* 46, no. 2 (April 2010): 113–35, https://doi.org/10.1017/S00322474 09008390.

30. "Statement of Apology for the Relocation of the Ahiarmiut," Government of Canada, January 22, 2019, https://www.rcaanc-cirnac.gc.ca/eng/154817025225 9/1548170273272.

31. Walter Strong and Jordan Konek, "Inuk Elder Recalls the Day Her Family Was Forced to Relocate, Nearly 70 Years Ago," CBC, February 2, 2019, https:// www.cbc.ca/news/canada/north/ahiarmiut-inuk-elder-forced-re location-1.5003380.

32. "'Dark Chapter in Our History': Federal Gov't Apologizes to Ahiarmiut for Forced Relocations," CBC, January 22, 2019, https://www.cbc.ca/news /canada/north/ahiarmiut-apology-federal-government-1.4986934.

33. *Community Climate Change Manual* (Nunavut: Arviat Aqqiumavvik Society, n.d.), 10.

34. Sarah Rogers, "Nunavut Man Dies in Kivalliq Polar Bear Attack," *Nunatsiaq News*, July 4, 2018, https://nunatsiaq.com/stories/article/65674nunavut_man _dies_in_polar_bear_attack/.

35. Rebecca Clare Harckham, "Defining and Servicing Mental Health in a Remote Northern Community," MSW dissertation, University of British Columbia, 2003, 51, https://doi.org/10.14288/1.0091109.

36. Katz, "With Old Traditions and New Tech, Young Inuit Chart Their Changing Landscape."

37. Julien Desrochers, "Aqqiumavvik Society HydroBlock Training," M2Ocean, August 2021, 19–21.

38. *Community Climate Change Manual*, 6.

39. Johannsdottir, Cook, and Arruda, "Systemic Risk of Cruise Ship Incidents from an Arctic and Insurance Perspective."

40. "Clipper Adventurer Cruise Ship Runs Aground in the Arctic," *Cruise Law News*, August 29, 2010, https://www.cruiselawnews.com/2010/08/articles /sinking/clipper-adventurer-cruise-ship-runs-aground-in-the-arctic/.

41. Wright and Baldauf, "Arctic Environment Preservation Through Grounding Avoidance," 77–78.

42. "Safety & Shipping Review 2021," Allianz Global Corporate & Specialty, August 2021, 9, https://www.agcs.allianz.com/news-and-insights/reports/shipping-safety/shipping-report.html.

43. Susan Nerberg, "I Returned to the Land of My Sámi Ancestors to Reclaim My Identity," *Broadview*, August 18, 2022, https://broadview.org/sami-colonization/.

44. John Noble Wilford, *The Mapmakers: The Story of the Great Pioneers in Cartography from Antiquity to the Space Age* (New York: Knopf, 1981), 167.

45. Kenn Harper, "The 'Boozy' Map of Nunavut," *Nunatsiaq News*, November 27, 2020, sec. Taissumani, https://nunatsiaq.com/stories/article/the-boozy-map-of-nunavut/.

46. "Pan Inuit Trails," Social Sciences and Humanities Research Council, http://www.paninuittrails.org/index.html?module=module.about.

47. Philip Steinberg, Jeremy Tasch, and Hannes Gerhardt, *Contesting the Arctic: Politics and Imaginaries in the Circumpolar North* (London: I. B. Tauris, 2015), 40.

48. Richard Kemeny, "Fight for the Arctic Ocean Is a Boon for Science," *Scientific American*, July 18, 2019.

49. Max Fisher, "Canada Just Enlisted Santa Claus in Its Effort to Control the Arctic," *Washington Post*, December 26, 2013, https://www.washingtonpost.com/news/worldviews/wp/2013/12/26/canada-just-enlisted-santa-claus-in-its-effort-to-control-the-arctic/.

50. Jane George, "Norway Wants Amundsen's Maud Back from Nunavut," *Nunatsiaq News*, May 16, 2011, https://nunatsiaq.com/stories/article/16557_norway_wants_amundsens_maud_back_from_nunavut/; Peter B. Campbell, "Opinion: Could Shipwrecks Lead the World to War?," *New York Times*, December 19, 2015, https://www.nytimes.com/2015/12/19/opinion/could-shipwrecks-lead-the-world-to-war.html.

51. Kemeny, "Fight for the Arctic Ocean Is a Boon for Science."

52. Government of Canada, "2016 Census—Census Subdivision of Arviat, HAM (Nunavut), https://www12.statcan.gc.ca/census-recensement/2016/as-sa/fogs-spg/Facts-csd-eng.cfm?LANG=Eng&GK=CSD&GC=6205015&TOPIC=4."

53. "New Global Survey Calls for Greater Coordination of Seabed Mapping Activities," Nippon Foundation-GEBCO Seabed 2030 Project, https://seabed2030.org/news/new-global-survey-calls-greater-coordination-seabed-mapping-activities.

54. Katrin Bennhold and Jim Tankersley, "Ukraine War's Latest Victim? The Fight Against Climate Change," *New York Times*, June 26, 2022, https://www.nytimes.com/2022/06/26/world/europe/g7-summit-ukraine-war-climate-change.html.

CHAPTER 8: THE ROBOT REVOLUTION AT SEA

1. "San Francisco to Hawaii Multibeam Mapping," Saildrone, https://www.Sail drone.com/missions/2021-surveyor-hawaii-mapping.

2. Brian Connon, "Who Is Going to Map the High Seas?," Hydro International, August 17, 2021, https://www.hydro-international.com/content/article/who -is-going-to-map-the-high-seas.

3. Scott Sistek, "Saildrone's Journey into Category 4 Hurricane Uncovers Clue into Rapidly Intensifying Storms," Fox News Network (Fox Weather, December 16, 2021), https://www.foxweather.com/weather-news/saildrones. -journey-into-category-4-hurricane-uncovers-clue-into-rapidly-intensifying -storms.

4. Dongxiao Zhang et al., "Comparing Air-Sea Flux Measurements from a New Unmanned Surface Vehicle and Proven Platforms During the SPURS-2 Field Campaign," *Oceanography* 32, no. 2 (June 2019): 122–33, https://doi .org/10.5670/oceanog.2019.220.

5. Susan Ryan, "Saildrone Closes $100 Million Series C Funding Round to Advance Ocean Intelligence Products," Saildrone, October 18, 2021, https:// www.saildrone.com/press-release/saildrone-announces-series-c-funding.

6. "USVs Complete Milestone Alaska Fisheries Survey," Saildrone, December 10, 2020, https://www.saildrone.com/news/usv-complete-milestone-alaska-poll ock-survey.

7. Denis Wood with John Fels, *The Power of Maps* (New York: Guilford Press, 1992), 7.

8. John Noble Wilford, *The Mapmakers: The Story of the Great Pioneers in Cartography from Antiquity to the Space Age* (New York: Knopf, 1981), 259.

9. Lloyd A. Brown, *The Story of Maps* (Boston: Little, Brown, 1949), 255.

10. Stephen Hall, *Mapping the Next Millennium: How Computer-Driven Cartography Is Revolutionizing the Face of Science* (New York: Random House, 1992), 384.

11. David Grann, *The Lost City of Z: A Legendary British Explorer's Deadly Quest to Uncover the Secrets of the Amazon* (London: Simon & Schuster, 2017), 51–52.

12. Wilford, *The Mapmakers*, 266.

13. Brown, *The Story of Maps*, 280.

14. Larry Mayer et al., "The Nippon Foundation-GEBCO Seabed 2030 Project: The Quest to See the World's Oceans Completely Mapped by 2030," *Geosciences* 8, no. 2 (February 8, 2018): 63, https://doi.org/10.3390/geo sciences8020063.

15. Brown, *The Story of Maps*, 300.

16. Wilford, *The Mapmakers*, 259.

17. Alastair Pearson et al., "Cartographic Ideals and Geopolitical Realities: International Maps of the World from the 1890s to the Present," *Canadian Geographer/Géographe Canadien* 50, no. 2 (June 2006): 149–76, https://doi.org/10.1111/j.0008-3658.2006.00133.x.

18. Jacqueline Carpine-Lancre et al., *History of GEBCO: 1903–2003* (Utrecht, Netherlands: GITC by Lemmer, 2003), 12.

19. Brown, *The Story of Maps*, 302.

20. Wilford, *The Mapmakers*, 236.

21. Ibid., 250.

22. Brown, *The Story of Maps*, 304.

23. Miles Harvey, *The Island of Lost Maps: A True Story of Cartographic Crime* (New York: Random House, 2000), 155.

24. Wilford, *The Mapmakers*, 251.

CHAPTER 9: BURIED HISTORY

1. Mackenzie E. Gerringer et al., "*Pseudoliparis swirei* sp. Nov.: A Newly-Discovered Hadal Snailfish (Scorpaeniformes: Liparidae) from the Mariana Trench," *Zootaxa* 4358, no. 1 (November 2017): 161–77, https://doi.org/10.11646/zootaxa.4358.1.7.

2. "Octopus Wonderland: Return to the Davidson Seamount," Nautilus Live, October 27, 2020, https://nautiluslive.org/video/2020/10/27/octopus-wonderland-return-davidson-seamount.

3. Sarah Durn, "The Northernmost Island in the World Was Just Discovered by Accident," *Atlas Obscura*, September 8, 2021, https://www.atlasobscura.com/articles/found-the-worlds-northernmost-island.

4. Henry Fountain, "At the Bottom of an Icy Sea, One of History's Great Wrecks Is Found," *New York Times*, March 9, 2022, updated July 13, 2022, https://www.nytimes.com/2022/03/09/climate/endurance-wreck-found-shackleton.html.

5. Neil Vigdor, "Sprawling Coral Reef Resembling Roses Is Discovered off Tahiti," *New York Times*, January 20, 2022, https://www.nytimes.com/2022/01/20/science/tahiti-coral-reef.html.

6. "Wrecks," UNESCO, https://web.archive.org/web/20220308003718/http://www.unesco.org/new/en/culture/themes/underwater-cultural-heritage/underwater-cultural-heritage/wrecks/.

7. Jay Bennett, "Less Than 1 Percent of Shipwrecks Have Been Explored," *Popular Mechanics*, January 18, 2016, https://www.popularmechanics.com/science/a19000/less-than-one-percent-worlds-shipwrecks-explored/.

8. Shawn Joy, "The Trouble with the Curve: Reevaluating the Gulf of Mexico

Sea-Level Curve," *Quaternary International* 523, no. 2 (July 2019), https://www.researchgate.net/publication/334566518_The_trouble_with_the_curve_Reevaluating_the_Gulf_of_Mexico_sea-level_curve.

9. Ole Grøn et al., "Acoustic Mapping of Submerged Stone Age Sites—A HALD Approach," *Remote Sensing* 13, no. 3 (January 2021): 445, https://doi.org/10.3390/rs13030445.

10. Megan Gannon, "7,000-Year-Old Native American Burial Site Found Underwater," *National Geographic*, February 28, 2018, https://www.nationalgeographic.com/adventure/article/florida-native-american-indian-burial-underwater.

11. Ibid.

12. Joy, "The Trouble with the Curve," 19.

13. A. Hooijer and R. Vernimmen, "Global LiDAR Land Elevation Data Reveal Greatest Sea-Level Rise Vulnerability in the Tropics," *Nature Communications* 12 (June 29, 2021): article 3592, https://doi.org/10.1038/s41467-021-23810-9.

14. "San Marcos de Apalache Historic State Park," Florida State Parks, https://stateparks.com/san_marcos_de_apalache_historic_state_park_in_florida.html.

15. Ole Grøn et al., "Detecting Human-Knapped Flint with Marine High-Resolution Reflection Seismics: A Preliminary Study of New Possibilities for Subsea Mapping of Submerged Stone Age Sites," *Underwater Technology* 35, no. 2 (July 2018): 35–49, https://doi.org/10.3723/ut.35.035.

16. Grøn et al., "Acoustic Mapping of Submerged Stone Age Sites," 12.

17. Ibid., 10.

18. Jennifer Raff, *Origin: A Genetic History of the Americas* (New York: Twelve, 2022), 73.

19. Thomas Curwen, "'Heroes, Villains, Charlatans, Enigmas': Why I Followed the Calico Story," *Los Angeles Times*, July 14, 2021, https://www.latimes.com/california/story/2021-07-14/heroes-villains-charlatans-enigmas-clueless-commentators-why-i-followed-the-story-of-calico.

20. Stefan Lovgren, "Clovis People Not First Americans, Study Shows," *National Geographic*, February 23, 2007, https://www.nationalgeographic.com/science/article/native-people-americans-clovis-news.

21. Stuart J. Fiedel, "Initial Human Colonization of the Americas: An Overview of the Issues and the Evidence," *Radiocarbon* 44, no. 2 (2002): 407–36, https://doi.org/10.1017/S0033822200031817.

22. Raff, *Origin*, 23.

23. Jessi J. Halligan et al., "Pre-Clovis Occupation 14,550 Years Ago at the Page-Ladson Site, Florida, and the Peopling of the Americas," *Science Advances* 2, no. 5 (May 13, 2016), https://doi.org/10.1126/sciadv.1600375.

24. "Monte Verde Archaeological Site," UNESCO World Heritage Centre, https://whc.unesco.org/en/tentativelists/1873/.

25. Ibid.

26. Raff, *Origin*, 78–79.

27. Halligan et al., "Pre-Clovis Occupation 14,550 Years Ago at the Page-Ladson Site, Florida, and the Peopling of the Americas."

28. Fiedel, "Initial Human Colonization of the Americas."

29. Jennifer Raff, "Rejecting the Solutrean Hypothesis: The First Peoples in the Americas Were Not from Europe," *Guardian*, February 21, 2018, http://www.theguardian.com/science/2018/feb/21/rejecting-the-solutrean-hypothesis-the-first-peoples-in-the-americas-were-not-from-europe.

30. Rob Diaz de Villegas, "Shells, Buried History, and the Apalachee Coastal Connection," *The WFSU Ecology Blog* (blog), May 29, 2012, https://blog.wfsu.org/blog-coastal-health/2012/05/shells-buried-history-and-the-apalachee-coastal-connection/.

31. "The Apalachees of Northwest Florida," Exploring Florida, http://fcit.usf.edu/Florida/lessons/apalach/apalach1.htm.

32. Barbara A. Purdy, *Florida's Prehistoric Stone Technology* (Gainsville: University Presses of Florida, 1981), xi.

33. Susan Schulten, *A History of America in 100 Maps* (Chicago: University of Chicago Press, 2018), 76–77.

34. George M. Cole and John E. Ladson, *The Wacissa Slave Canal* (Monticello, FL: Aucilla Research Institute, 2018), 23–24.

35. John Worth, "Rediscovering Pensacola's Lost Spanish Missions," paper presented at 65th Annual Meeting of the Southeastern Archaeological Conference, Charlotte, NC, November 15, 2008, https://www.academia.edu/2096954/Rediscovering_Pensacola_s_Lost_Spanish_Missions.

36. Dana Bowker Lee, "The Talimali Band of Apalachee," University of Louisiana Regional Folklife Program, accessed June 14, 2021, https://web.archive.org/web/20220407021802/https://www.nsula.edu/regionalfolklife/apalachee/Epilogue.html.

37. Purdy, *Florida's Prehistoric Stone Technology*, 1.

38. Raff, *Origin*, 16.

39. Joy, "The Trouble with the Curve."

40. "Operation Timucua: FWC Shuts Down Crime Ring Selling Priceless Florida Artifacts," *Woods 'n Water*, February 28, 2013, https://wnwpressrelease.wordpress.com/2013/02/28/operation-timucua-fwc-shuts-down-crime-ring-selling-priceless-florida-artifacts/.

41. Ben Montgomery, "North Florida Arrowhead Sting: What's the Point?," *Tampa Bay Times*, January 3, 2014, https://www.tampabay.com/features/humaninterest/north-florida-arrowhead-sting-whats-the-point/2159379/.

42. Daniel Ruth, "Ruth: Ridiculous 'Raiders of the Lost Artifacts,'" *Tampa Bay Times*, January 9, 2014, https://www.tampabay.com/opinion/columns/ruth-ridiculous-raiders-of-the-lost-artifacts/2160352/.

43. Rob Diaz de Villegas, "Amateur Archeologist vs. Looter: A Matter of Context?," *The WFSU Ecology Blog* (blog), November 6, 2015, https://blog.wfsu.org/blog-coastal-health/2015/11/amateur-archeologist-vs-looter-a-matter-of-context/.

44. Susan Ryan, "Saildrone's New Ocean Mapping HQ to Support Critical Florida Coastline Initiatives," Saildrone, March 2, 2022, https://www.saildrone.com/press-release/florida-ocean-mapping-hq-supports-critical-coastline-initiatives.

45. Robert Hanley, "Diving to Prove Indians Lived on Continental Shelf," *New York Times*, July 29, 2003, https://www.nytimes.com/2003/07/29/nyregion/diving-to-prove-indians-lived-on-continental-shelf.html.

46. *Archaeological Damage from Offshore Dredging: Recommendations for Pre-operational Surveys and Mitigation During Dredging to Avoid Adverse Impacts* (Herndon, VA: U.S. Department of Interior, February 2004), 19.

47. "Hurricanes," Florida Climate Center, https://climatecenter.fsu.edu/topics/hurricanes.

CHAPTER 10: MINING THE DEEP

1. Kyle Frishkorn, "Why the First Complete Map of the Ocean Floor Is Stirring Controversial Waters," *Smithsonian Magazine*, July 13, 2017, https://www.smithsonianmag.com/science-nature/first-complete-map-ocean-floor-stirring-controversial-waters-180963993/.

2. Stephen Hall, *Mapping the Next Millennium: How Computer-Driven Cartography Is Revolutionizing the Face of Science* (New York: Random House, 1992), 386.

3. Kendall R. Jones et al., "The Location and Protection Status of Earth's Diminishing Marine Wilderness," *Current Biology* 28, no. 15 (August 6, 2018): 2506–12.E3, https://doi.org/10.1016/j.cub.2018.06.010.

4. Holly J. Niner et al., "Deep-Sea Mining with No Net Loss of Biodiversity—An Impossible Aim," *Frontiers in Marine Science* 5 (March 2018), https://www.frontiersin.org/articles/10.3389/fmars.2018.00053.

5. Daniel O. B. Jones et al., "Biological Responses to Disturbance from Simulated Deep-Sea Polymetallic Nodule Mining," *PLOS ONE* 12, no. 2 (February 8, 2017): e0171750, https://doi.org/10.1371/journal.pone.0171750.

6. David Shukman, "Accident Leaves Deep Sea Mining Machine Stranded," *BBC News*, April 28, 2021, sec. Science & Environment, https://www.bbc.com/news/science-environment-56921773.

7. Helen Scales, *The Brilliant Abyss: Exploring the Majestic Hidden Life of the Deep Ocean, and the Looming Threat That Imperils It* (New York: Atlantic Monthly Press, 2021), 192.

8. Luc Cuyvers et al., *Deep Seabed Mining: A Rising Environmental Challenge* (Gland, Switzerland: IUCN and Gallifrey Foundation, 2018), 7.

9. John Childs, "Extraction in Four Dimensions: Time, Space and the Emerging Geo(-)Politics of Deep-Sea Mining," *Geopolitics* 25, no. 1 (January 2020): 189–213, https://doi.org/10.1080/14650045.2018.1465041.

10. Pradeep A. Singh, "The Two-Year Deadline to Complete the International Seabed Authority's Mining Code: Key Outstanding Matters That Still Need to Be Resolved," *Marine Policy* 134 (December 2021): article 104804, https://doi.org/10.1016/j.marpol.2021.104804.

11. Jenessa Duncombe, "The 2-Year Countdown to Deep-Sea Mining," *Eos*, January 24, 2022, https://eos.org/features/the-2-year-countdown-to-deep-sea-mining.

12. Jones et al., "The Location and Protection Status of Earth's Diminishing Marine Wilderness."

13. Nathalie Seddon et al., "Understanding the Value and Limits of Nature-Based Solutions to Climate Change and Other Global Challenges," *Philosophical Transactions of the Royal Society B: Biological Sciences* 375, no. 1794 (March 16, 2020): 20190120, https://doi.org/10.1098/rstb.2019.0120.

14. Museum exhibit, International Seabed Authority.

15. Monica Allen, "An Intellectual History of the Common Heritage of Mankind as Applied to the Oceans," thesis, University of Rhode Island, 1992, 108–9, https://digitalcommons.uri.edu/ma_etds/283.

16. Cuyvers et al., *Deep Seabed Mining*, 9.

17. Alan J. Jamieson and Thomas Linley, hosts, "Deep-Sea Mining Special," *The Deep-Sea Podcast*, episode 006, Armatus Oceanic, December 10, 2020, https://www.armatusoceanic.com/podcast/006-deep-sea-mining-special.

18. Jeffrey C. Drazen et al., "Midwater Ecosystems Must Be Considered When Evaluating Environmental Risks of Deep-Sea Mining," *Proceedings of the National Academy of Sciences* 117, no. 30 (July 28, 2020): 17455–60, https://doi.org/10.1073/pnas.2011914117.

19. Arlo Hemphill, "Greenpeace Intervention at the 26th Session of the International Seabed Authority," 26th Session of the International Seabed Authority, Kingston, Jamaica, December 7, 2021.

20. "Why the Rush? Seabed Mining in the Pacific Ocean," MiningWatch Canada, July 26, 2019, https://miningwatch.ca/publications/2019/7/17/why-rush-seabed-mining-pacific-ocean.

21. Cuyvers et al., *Deep Seabed Mining*, 35.

22. Ibid.

23. Elizabeth Kolbert, "Mining the Bottom of the Sea," *New Yorker*, December 26, 2021, https://www.newyorker.com/magazine/2022/01/03/mining-the-bottom-of-the-sea.

24. Ian Urbina, *The Outlaw Ocean: Journeys Across the Last Untamed Frontier* (New York: Alfred A. Knopf, 2019), xi.

25. Karen McVeigh, "Disappearances, Danger and Death: What Is Happening to Fishery Observers?," *Guardian*, May 22, 2020, https://www.theguardian.com/environment/2020/may/22/disappearances-danger-and-death-what-is-happening-to-fishery-observers.

26. Arlo Hemphill, "Greenpeace Intervention at the 26th Session of the International Seabed Authority," Greenpeace, December 7, 2021.

27. Jean Buttigieg, "Arvid Pardo: A Diplomat with a Mission," 2016, 13–28, https://www.um.edu.mt/library/oar/handle/123456789/14918.

28. Arvid Pardo, "Note Verbale: Request for the Inclusion of a Supplementary Item in the Agenda of the Twenty-Second Session," UN General Assembly, 22nd Session, New York, August 17, 1967, 7.

29. "William Wertenbaker, "Mining the Wealth of the Ocean Deep," *New York Times*, July 17, 1977, https://www.nytimes.com/1977/07/17/archives/mining-the-wealth-of-the-ocean-deep-multinational-companies-are.html.

30. Allen, "An Intellectual History of the Common Heritage of Mankind as Applied to the Oceans," 24.

31. Pardo, "Note Verbale."

32. Elaine Woo, "Arvid Pardo; Former U.N. Diplomat from Malta," *Los Angeles Times*, July 18, 1999, https://www.latimes.com/archives/la-xpm-1999-jul-18-me-57228-story.html.

33. Cuyvers et al., *Deep Seabed Mining*, 30–32.

34. Scales, *The Brilliant Abyss*, 181–82.

35. Marta Conde et al., "Mining Questions of 'What' and 'Who': Deepening Discussions of the Seabed for Future Policy and Governance," *Maritime Studies* 21 (September 2022): 327–38, https://doi.org/10.1007/s40152-022-00273-2.

36. Robert Kunzig, *Mapping the Deep: The Extraordinary Story of Ocean Science* (New York: W. W. Norton, 2000), 87.

37. Helen M. Rozwadowski, *Fathoming the Ocean: The Discovery and Exploration of the Deep Sea* (Cambridge, MA: Harvard University Press, 2005), 136–38.

38. Alan J. Jamieson and Paul H. Yancey, "On the Validity of the *Trieste* Flatfish: Dispelling the Myth," *Biological Bulletin* 222, no. 3 (June 2012): 171–75, https://doi.org/10.1086/BBLv222n3p171.

39. James Nestor, *Deep: Freediving, Renegade Science, and What the Ocean Tells Us About Ourselves* (New York: First Mariner Books, 2014), 208–9.

40. Morgan E. Visalli et al., "Data-Driven Approach for Highlighting Priority Areas for Protection in Marine Areas Beyond National Jurisdiction," *Marine Policy* 122 (December 2020): article 103927, https://doi.org/10.1016/j.marpol.2020.103927.

41. Scales, *The Brilliant Abyss*, 190–93.

42. Museum exhibit, International Seabed Authority.

43. Woo, "Arvid Pardo."

44. Allen, "An Intellectual History of the Common Heritage of Mankind as Applied to the Oceans," 96–101.

45. Aletta Mondre, "Down Under the Sea" (International Studies Association, 2017), http://web.isanet.org/Web/Conferences/HKU2017-s/Archive/212 b0e54-c916-42c7-866b-4894c4da6d84.pdf.

46. J. Brian Harley, *The New Nature of Maps: Essays in the History of Cartography* (Baltimore, MD: John Hopkins University Press, 2001).

47. Greg Stone, host, "Gerard Barron—CEO of DeepGreen: The Future of Energy Lies 4 Km Deep," *The Sea Has Many Voices* (podcast), episode 8, https://theseahasmanyvoices.com/project/gerard-barron-ceo-businesses/; Aryn Baker, "Seabed Mining May Solve Our Energy Crisis. But At What Cost?," *Time*, September 7, 2021, https://time.com/6094560/deep-sea-mining-environmental-costs-benefits/.

48. Baker, "Seabed Mining May Solve Our Energy Crisis."

49. Susan Schulten, *A History of America in 100 Maps* (Chicago: University of Chicago Press, 2018), 10.

50. Harley, *The New Nature of Maps: Essays in the History of Cartography*.

51. Scales, *The Brilliant Abyss*, 190.

52. Andrew Friedman, "After Chaotic Year, Seabed Mining Oversight Body Must Strengthen Policies," Pew, February 12, 2021, https://pew.org/2Ng KVwl.

53. Richard Fisher, "The Unseen Man-Made 'Tracks' on the Deep Ocean Floor," BBC Future, December 3, 2020, https://www.bbc.com/future/article/202012 02-deep-sea-mining-tracks-on-the-ocean-floor.

54. Drazen et al., "Midwater Ecosystems Must Be Considered When Evaluating Environmental Risks of Deep-Sea Mining."

55. Cuyvers et al., *Deep Seabed Mining*, 63–64.

56. Ibid.

57. David Shukman, "Accident Leaves Deep Sea Mining Machine Stranded," BBC News, April 28, 2021, https://www.bbc.com/news/science-environment-56921773.

58. Scales, *The Brilliant Abyss*, 192–98.

59. Cuyvers et al., *Deep Seabed Mining*, 64; Amy Maxmen, "Discovery of Vibrant Deep-Sea Life Prompts New Worries over Seabed Mining," *Nature* 561, no. 7724 (September 27, 2018): 443–44, https://doi.org/10.1038/d41586-018-06 771-w.

60. Drazen et al., "Midwater Ecosystems Must Be Considered When Evaluating Environmental Risks of Deep-Sea Mining."

61. Scales, *The Brilliant Abyss*, 199.

62. PBS, "Lessons from the Dust Bowl w/ Ken Burns (Live YouTube Event),"

YouTube, November 15, 2012, https://www.youtube.com/watch?v=g9GkNQa 5of8.

63. Ibid.

64. Jeffrey C. Drazen et al., "Midwater Ecosystems Must Be Considered When Evaluating Environmental Risks of Deep-Sea Mining," *Proceedings of the National Academy of Sciences* 117, no. 30 (July 28, 2020): 17458, https://doi.org /10.1073/pnas.2011914117.

65. Drazen et al., "Midwater Ecosystems Must Be Considered When Evaluating Environmental Risks of Deep-Sea Mining"; Diva J. Amon et al., "Assessment of Scientific Gaps Related to the Effective Environmental Management of Deep-Seabed Mining," *Marine Policy* 138 (April 2022): article 105006, https:// doi.org/10.1016/j.marpol.2022.105006.

66. Johnna Crider, "DeepGreen CEO Gerard Barron Opens Up About Deep-Green's Open Letter to BMW & Other Brands," CleanTechnica, April 14, 2021, https://cleantechnica.com/2021/04/14/deepgreen-ceo-gerard-barron -opens-up-about-deepgreens-open-letter-to-bmw-other-brands/.

67. K. A. Miller et al., "Challenging the Need for Deep Seabed Mining from the Perspective of Metal Demand, Biodiversity, Ecosystems Services, and Benefit Sharing," *Frontiers in Marine Science* 0 (2021), https://doi.org/10.3389 /fmars.2021.706161.

68. Beth N. Orcutt et al., "Impacts of Deep-Sea Mining on Microbial Ecosystem Services," *Limnology and Oceanography* 65, no. 7 (July 2020): 1489–1510, https://doi.org/10.1002/lno.11403.

69. Annie Leonard, Sian Owen, and Patrick Alley, "De-SPAC Merger of Sustainable Opportunities Acquisition Corp. (Ticker: SOAC; CIK: 0001798562) and DeepGreen Metals, Inc.," July 6, 2021, 5, https://savethehighseas.org/2021 /07/06/letter-to-sec-states-deep-sea-mining-company-has-misled-investors -ahead-of-going-public/.

70. Diva J. Amon et al., "Assessment of Scientific Gaps Related to the Effective Environmental Management of Deep-Seabed Mining," *Marine Policy* 138 (April 1, 2022): 11, https://doi.org/10.1016/j.marpol.2022.105006.

71. Louisa Casson, "Deep Trouble: The Murky World of the Deep Sea Mining Industry" (Greenpeace International, December 8, 2020), 10, https://www .greenpeace.org/international/publication/45835/deep-sea-mining-exploit ation.

72. "In Too Deep: What We Know, and Don't Know, About Deep Seabed Mining," World Wildlife Fund International, 2021, 4, https://files.worldwildlife.org/wwf cmsprod/files/Publication/file/1kgrh1yzmx_WWF_InTooDeep_What_we _know_and_dont_know_about_DeepSeabedMining_report_February_2021 .pdf?_ga=2.117983753.1063757461.1672107356-2084767261.167210 7354.

73. Diva Amon, Lisa A. Levin, and Natalie Andersen, "Undisturbed: The Deep

Ocean's Vital Role in Safeguarding Us from Crisis" (Oxford, United Kingdom: International Programme on the State of the Ocean, 2022), http://www .stateoftheocean.org/outreach/new-resources/.

74. Luise Heinrich et al., "Quantifying the Fuel Consumption, Greenhouse Gas Emissions and Air Pollution of a Potential Commercial Manganese Nodule Mining Operation," Marine Policy 114 (April 2020): article 103678, https:// doi.org/10.1016/j.marpol.2019.103678.

75. "Greenhouse Gas Equivalencies Calculator," US Environmental Protection Agency, March 2022, https://www.epa.gov/energy/greenhouse-gas-equivalen cies-calculator.

76. Kalolaine Fainu, "'Shark Calling': Locals Claim Ancient Custom Threatened by Seabed Mining," *Guardian*, September 30, 2021, https://www.theguardian .com/world/2021/sep/30/sharks-hiding-locals-claim-deep-sea-mining-off-papua -new-guinea-has-stirred-up-trouble.

77. Olive Heffernan, "Why a Landmark Treaty to Stop Ocean Biopiracy Could Stymie Research," *Nature* 580, no. 7801 (March 27, 2020): 20–22, https://doi. org/10.1038/d41586-020-00912-w.

78. Ibid.

79. Scales, *The Brilliant Abyss*, 130.

80. Ibid., 132–36.

81. *The Ocean Economy in 2030* (Paris: OECD, 2016), 200, https://doi.org/10.1787 /9789264251724-en.

82. Heffernan, "Why a Landmark Treaty to Stop Ocean Biopiracy Could Stymie Research."

83. Kolbert, "Mining the Bottom of the Sea."

84. Scales, *The Brilliant Abyss*, 220.

85. Cuyvers et al., *Deep Seabed Mining*, 55.

86. Karol Ilagan et al., "How the Rise of Electric Cars Endangers the 'Last Fron- tier' of the Philippines," NBC News, December 7, 2021, https://www.nbcnews .com/specials/rise-of-electric-cars-endangers-last-frontier-philippines/.

87. Ian Morse, "In Indonesia, a Tourism Village Holds Off a Nickel Mine—for Now," *Mongabay*, December 8, 2019, https://news.mongabay.com/2019/12/in -indonesia-a-tourism-village-holds-off-a-nickel-mine-for-now/.

88. Nick Rodway, "Nickel, Tesla and Two Decades of Environmental Activism: Q&A with Leader Raphaël Mapou," *Mongabay*, June 22, 2022, https://news .mongabay.com/2022/06/nickel-tesla-and-two-decades-of-environmental-act ivism-qa-with-leader-rapheal-mapou/.

89. Matt McFarland, "The Next Holy Grail for EVs: Batteries Free of Nickel and Cobalt," CNN, June 1, 2022, https://www.cnn.com/2022/06/01/cars/tesla-lfp -battery/index.html.

90. Andrew Thaler, "Has Pulling the Trigger Already Backfired?," *DSM Observer*,

August 26, 2021, https://dsmobserver.com/2021/08/has-pulling-the-trigger
-already-backfired/.

91. Kathryn Abigail Miller et al., "Challenging the Need for Deep Seabed Mining
from the Perspective of Metal Demand, Biodiversity, Ecosystems Services,
and Benefit Sharing," *Frontiers in Marine Science* 8 (July 29, 2021), https://
doi.org/10.3389/fmars.2021.706161; "Deep-Sea Mining: Who Stands to Benefit?," Deep Sea Conservation Coalition, fact sheet 6, February 2022, https://
savethehighseas.org/wp-content/uploads/2022/03/DSCC_FactSheet6
_DSM_WhoBenefits_4pp_Feb22.pdf.

92. Mehdi Remaoun, "Statement on Behalf of the African Group," 25th Session of
the Council of the International Seabed Authority, Kingston, Jamaica, February 25, 2019, https://isa.org.jm/files/files/documents/1-algeriaoboag
_finmodel.pdf.

93. Gerard Barron, "Address to ISA Council by Gerard Barron, CEO & Chairman of DeepGreen Metals, Member of the Nauru Delegation," ISA Council,
Kingston, Jamaica, February 27, 2019, https://www.isa.org.jm/files/files/doc
uments/nauru-gb.pdf.

94. Eric Lipton, "Secret Data, Tiny Islands and a Quest for Treasure on the Ocean
Floor," *New York Times*, August 29, 2022, https://www.nytimes.com
/2022/08/29/world/deep-sea-mining.html.

95. Khurshed Alam, "Letter Dated 28 June 2021 from the President of the ISA
Council," June 28, 2021, https://www.isa.org.jm/index.php/news/nauru-requests
-president-isa-council-complete-adoption-rules-regulations-and-procedures.

96. Chris Bryant, "$500 Million of SPAC Cash Vanishes Under the Sea,"
Bloomberg, September 13, 2021, https://www.bloomberg.com/opinion/articles
/2021-09-13/tmc-500-million-cash-shortfall-is-tale-of-spac-disappointment
-greenwashing.

97. Lipton, "Secret Data, Tiny Islands and a Quest for Treasure on the Ocean
Floor."

98. Louisa Casson, "Deep Trouble: The Murkey World of the Deep Sea Mining
Industry," Greenpeace, December 2020, 6–8, https://www.greenpeace.org
/static/planet4-international-stateless/c86ff110-pto-deep-trouble-report-final
-1.pdf.

99. Elin A. Thomas et al., "Assessing the Extinction Risk of Insular, Understudied
Marine Species," *Conservation Biology* 36, no. 2 (April 2022): e13854, https://
doi.org/10.1111/cobi.13854.

100. Cuyvers et al., *Deep Seabed Mining*, 42.

101. David Shukman, "The Secret on the Ocean Floor," BBC, February 19, 2018,
https://www.bbc.co.uk/news/resources/idt-sh/deep_sea_mining.

102. Colin Filer, Jennifer Gabriel, and Matthew G. Allen, "How PNG Lost US$120
Million and the Future of Deep-Sea Mining," *Devpolicy Blog*, April 28, 2020,
https://devpolicy.org/how-png-lost-us120-million-and-the
-future-of-deep-sea-mining-20200428/.

103. Ben Doherty, "Collapse of PNG Deep-Sea Mining Venture Sparks Calls for Moratorium," *Guardian*, September 15, 2019, https://www.theguardian.com /world/2019/sep/16/collapse-of-png-deep-sea-mining-venture-sparks-calls-for -moratorium.

104. "Mining's Tesla Moment: DeepGreen Harvests Clean Metals from the Sea-floor," Mining.com, June 15, 2017, https://www.mining.com/web/minings -tesla-moment-deepgreen-harvests-clean-metals-seafloor/.

105. Duncan Currie, "Deep Sea Conservation Coalition Intervention," paper presented at International Seabed Authority Council Meeting, Kingston, Jamaica, December 8, 2021, https://savethehighseas.org/isa-tracker/category /statements.

106. Casson, "Deep Trouble," 8.

107. Sue Farran, "COVID-19 Made Deep-Sea Mining More Tempting for Some Pacific Islands—This Could Be a Problem," *The Conversation*, June 14, 2021, https://theconversation.com/covid-19-made-deep-sea-mining-more -tempting-for-some-pacific-islands-this-could-be-a-problem-158550.

108. Kolbert, "Mining the Bottom of the Sea"; Elizabeth Kolbert, "The Deep Sea Is Filled with Treasure, but It Comes at a Price," *New Yorker*, June 6, 2021, https://www.newyorker.com/magazine/2021/06/21/the-deep-sea-is-filled -with-treasure-but-it-comes-at-a-price.

109. Kolbert, "The Deep Sea Is Filled with Treasure, but It Comes at a Price."

110. Bryant, "$500 Million of SPAC Cash Vanishes Under the Sea."

111. Elham Shabahat, "'Antithetical to Science': When Deep-Sea Research Meets Mining Interests," *Mongabay*, October 4, 2021, https://news.mongabay.com /2021/10/antithetical-to-science-when-deep-sea-research-meets-mining -interests/.

112. Rozwadowski, *Fathoming the Ocean*, 41.

113. Philomène A. Verlaan and David S. Cronan, "Origin and Variability of Resource-Grade Marine Ferromanganese Nodules and Crusts in the Pacific Ocean: A Review of Biogeochemical and Physical Controls," *Geochemistry* 82, no. 1 (April 2022): article 125741, https://doi.org/10.1016/j.chemer .2021.125741.

114. Shabahat, "'Antithetical to Science.'"

115. Todd Woody, "Do We Know Enough About the Deep Sea to Mine It?," *National Geographic*, July 24, 2019, https://www.nationalgeographic.com/envi ronment/article/do-we-know-enough-about-deep-sea-to-mine-it.

116. Shabahat, "'Antithetical to Science.'"

117. Justin Scheck, Eliot Brown, and Ben Foldy, "Environmental Investing Frenzy Stretches Meaning of 'Green,'" *Wall Street Journal*, June 24, 2021, https:// www.wsj.com/articles/environmental-investing-frenzy-stretches-meaning-of -green-11624554045.

118. David Edward Johnson, "Protecting the Lost City Hydrothermal Vent System:

All Is Not Lost, or Is It?," *Marine Policy* 107 (September 2019): article 103593, https://doi.org/10.1016/j.marpol.2019.103593.

119. "Marine Expert Statement Calling for a Pause to Deep-Sea Mining," Deep-Sea Mining Science Statement, https://www.seabedminingsciencestatement.org.

120. Casson, "Deep Trouble."

121. Johnson, "Protecting the Lost City Hydrothermal Vent System."

122. "Normand Energy Deep Sea Drilling Banner in San Diego," Greenpeace, March 31, 2021, https://media.greenpeace.org/archive/Normand-Energy-Deep-Sea-Drilling-Banner-in-San-Diego-27MDHUE66HH.html.

123. Cuyvers et al., *Deep Seabed Mining*, 44; Shukman, "Accident Leaves Deep Sea Mining Machine Stranded."

124. "Deep-Sea Mining: What Are the Alternatives?," Deep Sea Conservation Coalition, July 2021, https://savethehighseas.org/resources/publications/deep-sea-mining-what-are-the-alternatives.

125. Shukman, "Accident Leaves Deep Sea Mining Machine Stranded."

126. "The Metals Company Calls Video of Mining Waste Dumped into the Sea Misinformation as Stock Sinks," *MINING.COM* (blog), January 12, 2023, https://www.mining.com/the-metals-company-calls-video-of-mining-waste-dumped-into-the-sea-misinformation-as-stock-sinks/.

CHAPTER 11: TO THE BOTTOM AND BEYOND

1. Alan J. Jamieson et al., "Hadal Biodiversity, Habitats and Potential Chemosynthesis in the Java Trench, Eastern Indian Ocean," *Frontiers in Marine Science* 9 (March 8, 2022): article 856992, https://doi.org/10.3389/fmars.2022.856992.

2. Alan J. Jamieson and Michael Vecchione, "First in Situ Observation of Cephalopoda at Hadal Depths (Octopoda: Opisthoteuthidae: *Grimpoteuthis* sp.)," *Marine Biology* 167 (May 26, 2020): article 82, https://doi.org/10.1007/s00227-020-03701-1.

3. Josh Young, *Expedition Deep Ocean: The First Descent to the Bottom of the World's Oceans* (New York: Pegasus Books, 2020), 204.

4. Ibid., 191.

5. Cassandra Bongiovanni, Heather A. Stewart, and Alan J. Jamieson, "High-Resolution Multibeam Sonar Bathymetry of the Deepest Place in Each Ocean," *Geoscience Data Journal* 9, no. 1 (June 2022): 108–23, https://doi.org/10.1002/gdj3.122.

6. James V. Gardner, "U.S. Law of the Sea Cruises to Map Sections of the Mariana Trench and the Eastern and Southern Insular Margins of Guam and the Northern Mariana Islands," Center for Coastal and Ocean Mapping, University of New Hampshire, November 1, 2010, https://scholars.unh.edu/ccom/1255.

7. Young, *Expedition Deep Ocean*, 17.

8. Don Walsh, "Diving Deeper than Any Human Ever Dove," *Scientific American*, April 1, 2014, https://www.scientificamerican.com/article/diving-deeper-than-any-human-ever-dove/.

9. Ibid.

10. Ibid.

11. Rose Pastore, "John Steinbeck's 1966 Plea to Create a NASA for the Oceans," *Popular Science*, May 20, 2014, https://www.popsci.com/article/technology/john-steinbecks-1966-plea-create-nasa-oceans/.

12. Jyotika I. Vrimani, "Ocean vs Space: Exploration and the Quest to Inspire the Public," Marine Technology News, June 7, 2017, https://www.marinetechnologynews.com/news/ocean-space-exploration-quest-549183.

13. Alex Macon, "When SpaceX Rockets Take Flight (or Blow Up), LabPadre Is Watching," *TexasMonthly*, December 15, 2020, https://www.texasmonthly.com/news-politics/spacex-rockets-launch-labpadre-livestream/.

14. *FY 2020 Agency Financial Report*, NASA, https://www.nasa.gov/sites/default/files/atoms/files/nasa_fy2020_afr_508_compliance_v4.pdf.

15. Mike Read, "Virtual Conference: Industry Role in Seabed 2030," Marine Technology Society Virtual Symposia, June 11, 2020, https://register.gotowebinar.com/recording/3054056681389715723.

16. Paul Kiel and Jesse Eisinger, "How the IRS Was Gutted," ProPublica, December 18, 2018, https://www.propublica.org/article/how-the-irs-was-gutted.

17. Chris Isidore, "Elon Musk's US Tax Bill: $11 Billion. Tesla's: $0 | CNN Business," CNN, February 10, 2022, https://www.cnn.com/2022/02/10/investing/elon-musk-tesla-zero-tax-bill/index.html.

18. Kim McQuaid, "Selling the Space Age: NASA and Earth's Environment, 1958–1990," *Environment and History* 12, no. 2 (May 2006): 127–63, https://www.jstor.org/stable/20723571.

19. Carl Sagan, "The Gift of Apollo," *Parade*, January 11, 2014, https://parade.com/249407/carlsagan/the-gift-of-apollo/.

20. Maria Johansson et al., "Is Human Fear Affecting Public Willingness to Pay for the Management and Conservation of Large Carnivores?," *Society & Natural Resources* 25, no. 6 (June 2012): 610–20, https://doi.org/10.1080/08941920.2011.622734.

21. Alan J. Jamieson et al., "Fear and Loathing of the Deep Ocean: Why Don't People Care About the Deep Sea?," *ICES Journal of Marine Science* 78, no. 3 (July 2021): 797–809, https://doi.org/10.1093/icesjms/fsaa234.

22. Thoreau, quoted in Stephen Hall, *Mapping the Next Millennium: How Computer-Driven Cartography Is Revolutionizing the Face of Science* (New York: Random House, 1992), 399.

23. Thomas D. Linley et al., "Fishes of the Hadal Zone Including New Species, *in Situ* Observations and Depth Records of Liparidae," *Deep Sea Research Part I:*

Oceanographic Research Papers 114 (August 2016): 99–110, https://doi.org/10.1016/j.dsr.2016.05.003.

24. Cassandra Bongiovanni, Heather A. Stewart, and Alan J. Jamieson, "High-Resolution Multibeam Sonar Bathymetry of the Deepest Place in Each Ocean," *Geoscience Data Journal*, April 7, 2021, https://doi.org/10.1002/gdj3.122.

25. William J. Broad, "So You Think You Dove the Deepest? James Cameron Doesn't.," *New York Times*, September 16, 2019, sec. Science, https://www.nytimes.com/2019/09/16/science/ocean-sea-challenger-exploration-james-cameron.html.

26. "HOV DeepSea Challenger," Woods Hole Oceanographic Institution, https://www.whoi.edu/what-we-do/explore/underwater-vehicles/deepseachallenger/.

27. Vrimani, "Ocean vs Space."

28. "Five Deeps Expedition Is Complete After Historic Dive to the Bottom of the Arctic Ocean," Discovery, September 9, 2019, https://corporate.discovery.com/discovery-newsroom/five-deeps-expedition-is-complete-after-historic-dive-to-the-bottom-of-the-arctic-ocean/.

29. Guinness World Records, "Ocean's Deepest Point Conquered—Guinness World Records," YouTube, November 24, 2020, https://www.youtube.com/watch?v=ulIQ9_BB8KA.

30. Young, *Expedition Deep Ocean*, 273.

31. Adam Millward, "Earth's Tallest Mountain, Mauna Kea, Ascended for the First Time," Guinness World Records, December 29, 2021, https://www.guinnessworldrecords.com/news/2021/12/earths-tallest-mountain-mauna-kea-ascended-for-the-first-time-687258.

32. "1,058,522 Square Kilometers," The Measure of Things, Bluebulb Projects, http://www.bluebulbprojects.com/.

33. Amanda Holpuch, "New *Titanic* Footage Heralds Next Stage in Deep-Sea Tourism," *New York Times*, September 4, 2022, https://www.nytimes.com/2022/09/04/us/new-titanic-footage.html.

34. Randi Mann, "Why China Is Diving for Treasure in the Mariana Trench," Weather Network, November 11, 2020, https://www.theweathernetwork.com/ca/news/article/why-china-is-diving-for-treasure-in-the-mariana-trench.

35. Michael Verdon, "Meet the Modern-Day Adventurer Who Explores Space and Sea," *Robb Report*, May 19, 2022, https://robbreport.com/motors/aviation/victor-vescovo-modern-day-adventurer-1234680636/.

EPILOGUE

1. "Seabed 2030 Announces Increase in Ocean Data Equating to the Size of Europe and Major New Partnership at UN Ocean Conference," Nippon Foundation-GEBCO Seabed 2030 Project, accessed August 26, 2022, https://

seabed2030.org/news/seabed-2030-announces-increase-ocean-data
-equating-size-europe-and-major-new-partnership-un.

2. Stephen Hall, *Mapping the Next Millennium: How Computer-Driven Cartography Is Revolutionizing the Face of Science* (New York: Random House, 1992), 385.

3. John Noble Wilford, *The Mapmakers: The Story of the Great Pioneers in Cartography from Antiquity to the Space Age* (New York: Alfred A. Knopf, 1981), 313–22.

4. Ibid., 287–90.

5. "Brazil: Amazon Sees Worst Deforestation Levels in 15 Years," BBC News, November 19, 2021, https://www.bbc.com/news/world-latin-america -59341770.

6. Morgan E. Visalli et al., "Data-Driven Approach for Highlighting Priority Areas for Protection in Marine Areas Beyond National Jurisdiction," *Marine Policy* 122 (December 2020): article 103927, https://doi.org/10.1016/j.marpol .2020.103927.

RECOMMENDED READING

Atwood, Roger. *Stealing History: Tomb Raiders, Smugglers, and the Looting of the Ancient World.* New York: St. Martin's Press, 2004.

Brown, Lloyd A. *The Story of Maps.* Boston: Little, Brown, 1949.

Grann, David. *The Lost City of Z: A Legendary British Explorer's Deadly Quest to Uncover the Secrets of the Amazon.* London: Simon & Schuster Ltd., 2017.

Harley, J. Brian. *The New Nature of Maps: Essays in the History of Cartography.* Baltimore, MD: John Hopkins University Press, 2001.

Hayton, Bill. *The South China Sea: The Struggle for Power in Asia.* New Haven, CT: Yale University Press, 2014.

Kunzig, Robert. *Mapping the Deep: The Extraordinary Story of Ocean Science.* New York: W. W. Norton, 2000.

Menard, Henry William. *The Ocean of Truth: A Personal History of Global Tectonics.* Princeton, NJ: Princeton University Press, 1986.

Oreskes, Naomi. *Plate Tectonics: An Insider's History of the Modern Theory of the Earth.* Boulder, CO: Westview Press, 2001.

———. *Science on a Mission: How Military Funding Shaped What We Do and Don't Know About the Ocean.* Chicago: University of Chicago Press, 2021.

Raff, Jennifer. *Origin: A Genetic History of the Americas.* New York: Twelve, 2022.

Rozwadowski, Helen M. *Fathoming the Ocean: The Discovery and Exploration of the Deep Sea.* Cambridge, MA: Harvard University Press, 2005.

Scales, Helen. *The Brilliant Abyss: Exploring the Majestic Hidden Life of the Deep Ocean, and the Looming Threat That Imperils It.* New York: Atlantic Monthly Press, 2021.

Schulten, Susan. *A History of America in 100 Maps.* Chicago: University of Chicago Press, 2018.

Starosielski, Nicole. *The Undersea Network.* Durham, NC: Duke University Press, 2015.

Steinberg, Philip, Jeremy Tasch, and Hannes Gerhardt. *Contesting the Arctic: Politics and Imaginaries in the Circumpolar North.* London: I. B. Tauris, 2015.

Tester, Frank, and Peter Kulchyski. *Tammarniit (Mistakes): Inuit Relocation in the Eastern Arctic, 1939–63.* Vancouver: University of British Columbia Press, 2011.

Wilford, John Noble. *The Mapmakers: The Story of the Great Pioneers in Cartography from Antiquity to the Space Age.* New York: Knopf, 1981.

Wood, Denis, with John Fels. *The Power of Maps.* New York: Guilford Press, 1992.

Young, Josh. *Expedition Deep Ocean: The First Descent to the Bottom of the World's Oceans.* New York: Pegasus Books, 2020.

INDEX